DESIGNING AND
BUILDING MODEL RC
WARSHIPS

DESIGNING AND BUILDING MODEL RC
WARSHIPS

GLYNN GUEST

THE CROWOOD PRESS

Contents

Chapter One

Introduction

The urge to create something new and original is a common one and people usually find this to be a rewarding challenge. If you are lucky then you may have a job that involves this sort of work and actually pays you for doing something you enjoy. However, most of us are not so lucky, but there is always the option to take up a creative hobby.

Building reduced-scale models has a long history. Our ancestors, as soon as they learned to shape materials, started to make models of things like figures and animals. This could have just remained an artistic activity but they also had an educational value and became toys for the young to learn about the world. It was not hard to recognise that they could also be useful in science, technology and engineering. The scale may have been reduced but they could often prove that an idea or technique would work or not. This would be a much quicker and cheaper way to find and solve any problems before starting on the real thing. This is especially true of anything dynamic, such as vehicles, where small changes can have significant effects on the safe, reliable and economic operation of the full size.

A static model, no matter how good it looks, can lack the appeal of one that can move in a realistic fashion. The earliest model boats that could actually sail like the real thing would have been based on vessels with sails, wind being willing to move boats no matter their size. The advent of power sources small enough to fit inside a model boat, like clockwork, steam and electric motors, opened up the potential to make working models of virtually every type of vessel. Radio control then allowed these models to be fully under the command of the operator.

Now you can buy a model boat 'ready to run' – just charge the batteries, drop it in the water and sail off. This can give you pride of ownership and operation but not much of the creative satisfaction we first talked about. Many people build a model from a commercial kit or a published plan. This certainly needs involvement that is more personal and hopefully generates greater satisfaction of the 'I made it' kind. However, there is often the desire to build a model for which no commercial kit or plan is available, a truly original piece of work.

Models based on warships are often regarded as 'difficult' things to build and to safely operate. Some people build warship models that are of 'museum quality' with every little detail visible, which can take vast amounts of time, skill and money to complete. This is something that not everyone has the inclination to commit to, in what should be a relaxing hobby. It is possible to strike a balance between these demands by adopting the 'stand off scale' approach. The model is simplified, but not to the point where it spoils the appearance and sailing performance. Because it is intended to be viewed when sailing, every little detail that could not be seen under these conditions anyway, can be omitted. The result should be a model that you can honestly say 'I built it all myself'.

Readers might have some relevant experience outside the area of model boating. This would mean that some sections in this book overlap with already developed skills and experience. For example, you may have operated radio-controlled (RC) cars or aircraft models, but it still might be worth reading sections you know something about. Model boats can have their own unique peculiarities – a bit like the people who build and operate them.

RESEARCH

This term can sound intimidating and put otherwise talented and able people off the idea of making a model boat with any pretentions towards being a representation of a full-size vessel. In fact, some characters in this hobby will take great delight in boasting about the amount of time, money and effort they had to spend researching their 'masterpieces' before they could actually start to build it. It might be best to leave them in their world, as we are aiming to produce a stand off scale (SOS) model that, with much less heartache (physical, mental and financial), will result in a model that looks the part and you can enjoy sailing without worrying about the investment.

TWENTY YEARS AGO

Things are somewhat easier now with the vast amount of information available via the Internet, not so when the idea came to make a model based on the Peruvian turret-ram 'Huascar'. It

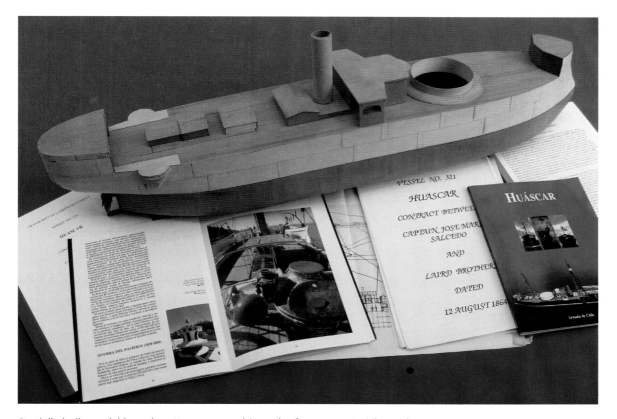

Partially built model based on Huascar *warship and reference materials used.*

had a very different appearance along with an intriguing history. She was built in England in the 1860s for the Peruvian navy; at that time, many South American countries were trying to establish their independence from Spain. While the *Huascar* arrived too late to join the conflict with Spanish forces, its crew were involved in a revolution in Peru and became somewhat piratical. This resulted in two Royal Navy vessels having an inconclusive battle with it. Later, during hostilities with neighbouring Chile, the *Huascar* was captured and incorporated into the Chilean navy, where it remains to this day. Luckily, in 1971 she was restored, which was quite a task as there was, understandably, little reference material to work with. She has now become a Chilean national monument moored at Talcahuano, much like HMS *Victory* is in Portsmouth.

Designing a model of *Huascar* created the same problem that the restorers faced – the lack of sufficient details. The best found were simple sketch drawings of the outline, some internal details and a basic deck plan. This showed that a viable model could be made, but left many puzzling areas. Historical references only had illustrations of limited detail, not helped by the numerous changes that it had undergone in its long history. For example, it had been fitted with an extensive outfit of sails and rigging during the delivery voyage from England, wind power being essential to supplement the modest 300 tons of coal it could carry for the steam engines.

All this was rather frustrating; the real thing existed several thousand miles away, but my hobby budget would not stretch to a visit to take all the photos needed to fill these gaps. In an attempt to locate more information, I wrote to the Chilean Embassy in London, explaining my problem and asking if they could point me in the direction of better information. An encouragingly prompt response came with the promise of further help and this arrived within a couple of months. The Chilean Naval Mission had obtained two tourist guides for visitors to the *Huascar*. These were in Spanish, but no translations were needed as they were beautifully illustrated and answered most of my questions.

A LITTLE BETTER NOW

When I came to build the Type 23 frigate there was much more information available. Collecting information was started a couple of years before any building commenced. As a result, a folder contained several magazine articles and a couple of Royal Navy recruitment booklets. The latter might be an easily overlooked source of very valuable photographs, freely available, but if you go into a recruiting office, modellers that are more senior might first want to explain that they are not thinking of joining the Navy!

The Internet threw up some highly detailed plans of these vessels but with a price to match. As an SOS model was being contemplated, buying these plans did not seem justified, as they would possibly double the building cost of the model. A few simpler drawings and photos on the Internet were sufficient to supply the basic proportions and layout of the Type 23. If there is a moral in this experience, it is not to rush into a project, but to start collecting useful material for as long as possible and keep your eyes open all the time. An example of this was the chance discovery of a book (*The Model Ship* by Norman Napier Boyd – ISBN 1 85149 327 1) that contained photographs of Type 23 models.

GOOD BOOKS

It might seem that the Internet could supply all the information you might need to build a good model. Just put the name of a vessel or class/type into a search engine and a flood of potential sources might emerge. You need, however, to exercise a degree of critical judgement as some could be limited in accuracy and value. Nevertheless, it is often possible with the right search criteria, to find a 'gem' or two.

Completed model of Type 23 frigate along with the references used.

Books might be considered as 'old hat' to many, but a book should have gone through many cycles of rewriting, checking and editing before it reaches your hands. This ought to mean that its content can be trusted to be accurate. Luckily, there are many books written on the subject of warships and ships in general. Possibly a touch ironically, the Internet can be very good at locating just the right book to meet your needs when planning to build a model. Some websites give you a preview of a book's content, which can help, especially if you are looking for features like scale drawings. Discount bookstores can sometimes be a treasure trove for reference books; there are a few that I cannot pass without going in, and over the years I have found some unbelievable bargains.

There is also the public library system to consider. While a local branch may only have a limited stock, if you become a member, then you can access the on-line catalogue, with the potential to borrow a book from anywhere in the country. Using an author's name or subject matter can locate a surprising range of books. This is especially valuable if the book you want is out of print, rare or perhaps too expensive to buy. My small local branch library has yet to let me down when ordering such items. Finally, books and book tokens make good birthday and Christmas presents, perhaps more welcome than another item of clothing.

A bibliography is included and is based on the books I have in my possession. It is not intended to

Discount book collection used for some models.

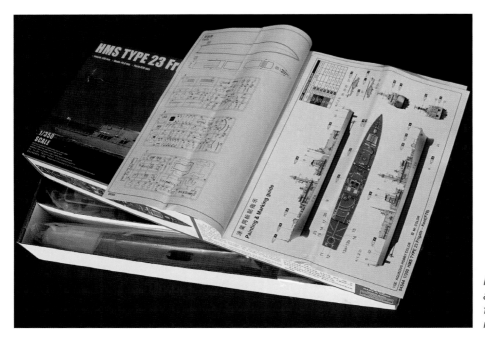

Plastic kits can be a good reference for larger working models.

be encyclopaedic, but to give you an idea of what is available. Also, using these authors' names when searching can lead you to much useful material.

PLASTIC INSPIRATION

Another possible item to help with designing a working model could be one of the small plastic construction kits. Even at a much smaller scale such as 1/600, the manufacturers take great care to achieve accurate shapes. This can be used to get the correct proportions for a larger model and illustrate the sometimes complex areas on the full-size vessels.

Some kit instructions also include useful drawings and painting details. These, combined with good photographs of the real things, are often enough to make a realistic working SOS model.

Good enough for the majority but, of course, never enough for the aesthetes and purists that may lurk around the lake, but then nothing ever is.

TELEVISION AND FILMS

Many would consider these two media to be at best entertaining diversions (at worst mind dumbing), but they can occasionally be a very useful source of information. This was brought home to me when I chanced upon a TV programme about chaplains in the Royal Navy. It was based upon a chaplain's first posting on board a ship, which just happened to be a Type 23 frigate. While an interesting subject in itself, many of the background images were a godsend for anyone building a model of these vessels. Alas, a little too late for me as I'd more or less completed my model!

Chapter Two

Size and Scale

Having decided which vessel they plan to base their model upon, an inexperienced person might then rush into designing and building it. This can lead to problems, which might not become apparent until much later, possibly just as the model is being prepared for its first sailing.

SCALE MATTERS

Scale in this sense simply refers to the ratio of the model's size compared with the full-size item. It is usually quoted as a fraction, for example 1/100 means all items on the model have been reduced to a linear one-hundredth of the full size.

You could build a model to any size you fancy, but there are some scales that have become popular with modellers and manufacturers. Even when building your own original creation, it can sometimes be handy to use suitable commercial items. Table 1 lists a range of popular scales and the modelling areas in which they are commonly found; this should not stop you 'borrowing' them for your model.

Plastic aircraft kits of the right type and scale have saved me lots of anguish when outfitting the flight decks of aircraft-carrier models. The thought of making from scratch such numerous and very obvious things that have to look identical does not bear thinking about.

Table 1 Common scales used in models.

Scale	Model Type
1/24	G Gauge Railway
1/32	Military + Plastic Kits
1/35	Military + Plastic Kits
1/48	O Gauge Railway + Plastic Kits
1/64	Die Cast Cars
1/72	Plastic Kits
1/87	HO/OO Gauge Railway
1/96	Ships
1/144	Plastic Kits
1/160	N Gauge Railway
1/192	Ships
1/220	Z Gauge Railway

Even a little bit of lateral thought can be handy, such as fashion jewellery, which can make anchor chains, not perfect maybe but close enough and painless. There are also many small businesses that can supply scale details and fittings for this hobby, usually at these common scales. Therefore, choosing a scale for your model, even if it is an original design, that matches or is close to one of the common commercial scales, does make sense.

These scales might seem to have 'odd' numbers like 1/96 rather than the neater looking 1/100.

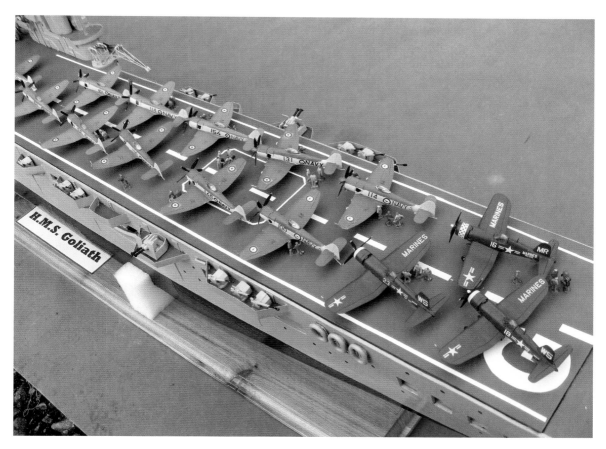

Well-populated aircraft carrier flight deck.

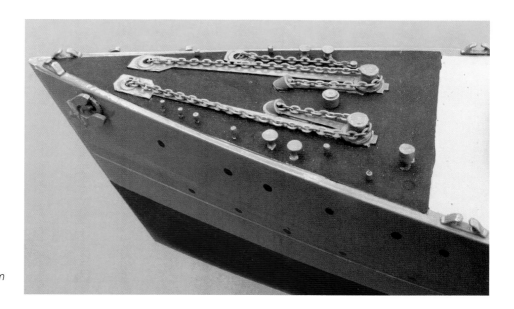

Anchor chain from cheap jewellery.

This is due to their origins, which might have a practical rather than a coldly logical basis. The multiples of twelve are related to the imperial units of twelve inches to the foot. Thus, a scale of 1/48 can be expressed as one inch equals four feet or 48 inches. To be honest, you can usually

Slightly over-scale ladders.

mix near scales without causing visual offence. A personal dislike is making ladders – the model railroad scales can provide suitable items. N gauge railway ladders (1/160) have often adorned my 1/144 scale models and no one has yet noticed this scale mismatch!

A further consideration could be if you plan to build more than one type of vessel to a common scale. Coming into this hobby via model aircraft made me familiar with building models using balsa sheets. It seemed sensible to continue with this material and so early models based on destroyers were sized to fit the standard balsa sheet lengths. This worked out to be around the imperial scale of one model inch equalling twelve full-size feet or 1/144. This produced practical models that performed well, not too expensive and, as found out after building quite a few, easy to store. Only later did it become apparent that this scale would allow me to build a range of warship models.

Table 2 shows the effect of scale on three common types of warship. It ought to be clear

Series of warship models built to a common scale.

that the 1/96 scale makes a practical destroyer model, the cruiser is becoming a handful, but the battleship would be quite a challenge to transport, let alone getting it in and out of the water. At scale 1/192, things are turned around with a manageable battleship model, the cruiser is fine but the destroyer has become a featherweight – possible, but quite a challenge, unless you accept sailing in only calm conditions. When built in 1/144 scale, the destroyer is a useful size and you do not need to be a weightlifter to cope with the battleship. This admittedly fortuitous discovery has been proven with models based on cruisers and aircraft-carriers built in this scale, but I have yet to build a battleship model.

STABILITY
(OR HOW TO REMAIN UPRIGHT)

It is worth spending a little time on how boats, model and full-size, manage to float and to know which way up they should be floating. Not essential knowledge, until you have problems.

The model's operating weight (often-termed displacement) is perhaps the best thing to calculate first. Boats, model and full-size, float due to Archimedes' principle, which is, as the hull descends into water, it 'pushes' (displaces) water out of the way. The water pushes back with force equal to the weight of water displaced (up-thrust). Hopefully, the weight of water displaced will match the model's weight before it submerges. This is a stable position: push the model down and the up-thrust exceeds the model's weight, so it rises back to the equilibrium position. This explains why models sink lower in the water as more weight is added to them.

It also shows why models are stable, even if there is apparently more of the model above than below the waterline around the hull. Two 'centres' are needed to explain this. The first is the centre of gravity, the point inside the model around which the total weight acts. The centre of buoyancy is the point inside the hull where the up-thrust force acts. It is the position of the displaced water's own centre of gravity. When the model is trimmed to float level, then the upwards' and downwards' forces are directly in line and in equilibrium. This is the same situation as a 'see-saw' being balanced on its central pivot.

Should the bows of the model be pushed downwards, then more of the forward section of the hull is immersed. This moves the centre of buoyancy forwards and now the up-thrust and

Table 2 Comparison of the effects of differing scales on the size of a model.

Scale	1/192	1/144	1/96
Destroyer 360ft 2000T	22in 1lb (0.5kg)	30in 2.5lb (1.1kg)	45in 8lb (3.7kg)
Cruiser 600ft 10,000T	38in 5lb (2.3kg)	50in 12lb (5.5kg)	75in 40lb (18kg)
Battleship 720ft 35,000T	45in 18lb (8.2kg)	60in 45lb (21kg)	90in 140lb (64kg)

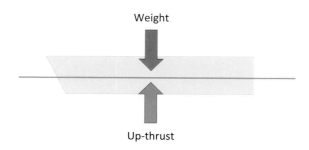

Why boats (model and full size) float.

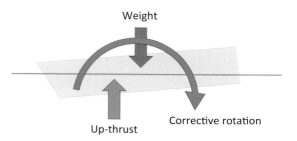

Movement of centre of buoyancy corrects bows going downwards.

weight, which should not move, act to rotate the model back to the level state.

Transverse stability can be a problem and sometimes you can witness someone's pride and joy rolling from side to side, as it sails along, possibly with any spectators wickedly watching to see if it turns upside-down. A test that I give to every new model is to roll it by pushing down on the edge of the deck until it is at the level of the water. Upon releasing the model, it ought to spring smartly upright, probably oscillating a few cycles, before ending back upright. This has always been proof that the model has adequate transverse stability and, unless I sail it in stupidly rough conditions, no problems will occur.

Most people will realise that this stability requires the model's centre of gravity to be low, but are surprised that it can still be above the centre of buoyancy. This sounds like an unstable situation as, when rolled, weight pushes down, while up-thrust pushes up, and it seems like it ought to carry on rotating.

The important thing is that rolling the model immerses more of the hull on the lower side, while reducing it on the raised side. This moves the centre of buoyancy to the lower side and the up-thrust, combined with the weight force, ought to act to restore the model to the upright position.

It is worth mentioning two terms that are important for transverse stability: the metacentre and the metacentric height. The former is the notional point on the vertical centreline of the hull about

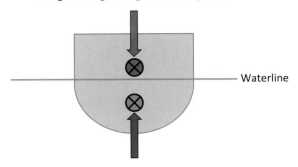

Weight acting through centre of gravity

Waterline

Up-thrust acting through centre of buoyancy

At rest with the centres in line.

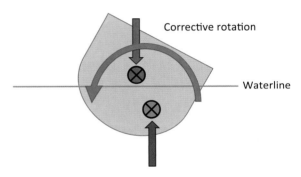

Corrective rotation

Waterline

Centre of buoyancy moves to low side and the out-of-line forces roll the hull back upright.

which the centre of buoyancy appears to swing as the model rolls. The metacentric height is the distance that this metacentre is above the centre of gravity of the hull. This has a positive value if the metacentre is above the centre of gravity, positive being stable and good. If the metacentric is below, a negative value is used because it is unstable. The metacentric height of a model can be found from a simple inclining experiment, but successful operation of a model does not need the knowledge of its value. However, what is desirable is an understanding of its existence, plus ensuring that your model has an adequate positive value.

All of this goes to show that for a given depth of hull immersion, a wider beam of model will be

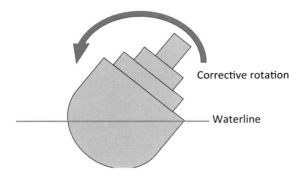

Corrective rotation

Waterline

Stability tested by heeling the hull.

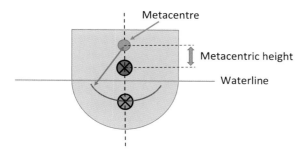

Centre of buoyancy swings about the metacentre.

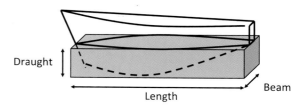

Block coefficient – the fraction of a rectangular block that the hull just fits inside.

more stable as the centre of buoyancy will move further to the low side for the same angle of roll. However, for all models it will pay to keep the centre of gravity low, everything above the waterline as light as possible and internal weights low.

PRECAUTIONARY CALCULATIONS

Some people avoid mathematics like the plague, but a few simple calculations before starting to make any working model of your own design can save a lot of heartache. For example, will it float correctly, can all the planned internal items fit inside the hull, and can it even be transported to the lake and launched safely? The most extreme size problem can be a model built indoors, which is then found to be too large to move outdoors without some unwelcome structural changes to the model and maybe house.

You could be put off by trying to use techniques used in full-size boat and shipbuilding, where accurate results are vital and demand complex methods. With our models we can relax a little and simply settle for a reasonable estimate using the 'block coefficient' and water density. You take the length, beam and draught of your hull; multiply them together to give you a 'block' of water that the underwater part of the hull will fit into. The block coefficient is simply the fraction of that block that the hull will actually occupy. For slim vessels (like fast warships) this fraction might only be a half (0.5) but for 'tubbier' models it would be

more. Using this figure multiplied by the volume of the 'block' of water, you have the hull volume, which if then multiplied by the density of water (appropriate units of course!), gives you an estimate of the model's weight.

Experience with many models based on displacement vessels has shown that if using inches to measure the hull and get the value of the 'block' in cubic inches, then multiplying this by $3/8$ (0.375) it will produce, if not a perfect value, at least a good workable estimate of the model's probable weight in ounces. The $3/8$ (0.375) being a product of an average block coefficient and the density of water. If using metric units like centimetres and grams, then a different number would be needed; my rusty maths suggests it ought to be around 0.67 with these units.

This simple weight calculation proved invaluable with the *Huascar* model, as my warship models are often built at 1/144 scale. However, at 1/144 scale, a *Huascar* model's length would be 40cm (16in) with a weight of 0.6kg (22oz), which seemed to be uncomfortably small. Going up to the next popular scale of 1/96 gave a more promising length of 60cm (24in) and a weight of 2kg (72oz). But, 1/72 scale with a model length of 81cm (32in) and 5kg (174oz) looked to be a much safer bet and more impressive.

The Type 23 model's scale was already fixed at 1/144, so that it would be compatible with my other warship models. This gave me a model length of around 91cm (36in) and a 1.8kg (64oz) weight. This was much more comfortable and well within my successful experiences.

These simple calculations before building also allow you to 'play' with the effects of small changes. You may feel that the planned model length is fine (fits in the car?), but wish for a little more internal space or greater weight. With an SOS type of model, you can often get away with small changes to a model's beam and underwater draught. A quick piece of arithmetic can show you how much can be gained by these changes before committing yourself.

SUMS FOR SPEED

Powering a new model is often a problem if you want to achieve a reasonable scale-like performance. Those who feel the need to impress or maybe intimidate others, will no doubt just stick in the biggest motor and as many shovelfuls of batteries as possible.

Many people are puzzled that when operating a scale-model boat, if you just multiply the full-size vessel's top speed by the scale, it can look way too slow. An example is something like a ship with a top speed of say 24 knots (kn) (a knot is one nautical mile per hour or about 51cm/s or 20in/s). This means the vessel moves forwards around 1,200cm/s (480in/s). If a model is built to 1/100 scale and you multiply these values by this scale, it will cover 12cm/s (less than 5in/s). This would produce the same time for both the full-size and model to pass a fixed point. In many people's eyes, this is acceptable, but when sailing at a distance away from any fixed reference points, it can look slow. The model fails to create a wave pattern that matches what you would expect the full-size vessel to generate at 24kn. You may have to look for a second or two before it is clear that the model is actually moving at all.

This problem was quickly recognised by film-makers when attempting to make realistic action scenes using scale-model boats (a lot cheaper than using the real things, especially if they wanted to sink them). They ran the cameras at a faster speed than normal, so that when the film was played

back, the wave and model motions looked much better. We can do the same by running our models at the 'dynamic scale' speeds, which produce the same wave patterns. This is based on the work of the nineteenth-century engineer William Froude and others when developing the theory of scale-model testing in naval architecture.

To produce the same visual wave pattern in the model as the full-size would at any speed, you scale the model's speed down by the square root of the scale used. In this example the 1/100 scale model needs to be travelling at 1/10 scale speed because:

$$\frac{1}{10} \times \frac{1}{10} = \frac{1}{100}$$

of the full-size speed. Therefore, to look like its speed matches 24 full-size knots, the model has to move at 1.2m/s (4ft/s). This might seem to be rather fast, but when sailing away from any fixed points that confuse the situation, it can look surprisingly realistic.

The *Huascar* model was a good example of the need to do these sums. The full-size had a top speed of around 10kn, which corresponds to about 5.2m/s (17ft/s). To find the model speed that creates the same wave pattern (that is, dynamically similar) you multiply this by the square root of the model's scale (1/72). So, 1/72 = 0.014 and the square root of this is 0.118.

The dynamically similar speed of the *Huascar* model will be

17 × 0.118 = 2ft/s (61cm/s)

This is quite a modest speed and no problems were anticipated.

The Type 23 had the same calculations undertaken, this time with a full-size speed of 30kn (maybe a shade more as the military often publically understate the performance of their stuff) with a scale of 1/144.

Square root of 1/144 is 1/12 or 0.083

30kn = 15m/s (50ft/s)

so dynamic scale speed of the model is:

$$50 \times 0.083 = 4.2\text{ft/s (1.3m/s)}$$

A reasonable figure and not one to worry about.

PONDERING POWER

Another item worth thinking about is providing the model with sufficient power to perform in a realistic manner. Nothing looks worse than, say, a model based on a destroyer (the greyhound of the sea!) wallowing along at the proverbial snail's pace. Equally silly is the sight of an overpowered warship model racing around with its bows clear of the water and the stern below the water level.

It is possible to use the power produced by the engines of the full-size vessel and a scaling factor to estimate the power needed to drive your model at a realistic speed. However, having collected the performance data for many models, I came up with a simple way to estimate the probable power to achieve the performance and handling characteristics needed for a new model. It is not perfect and assumes the correct installation and operation of the model's power system, but as Americans are wont to say, 'It usually gets you in the right ballpark'.

The model's behaviour when sailing can be estimated from the ratio of the input power to its weight. Naturally, the greater is this ratio, the faster in speed and steering response will it become. Table 3 breaks these down into four categories.

The 'steady' category has the lowest ratio and usually produces modest performance. While very relaxing to sail, it should never run into trouble, it is perhaps not suited to models based upon most warships. The 'handy' ratio is more promising, still not difficult to sail, but better performance. Moving up to 'lively' and you can have something better matching what a model based upon a destroyer, frigate or cruiser would be expected to

An unhappy and potentially unsafe overpowered model.

The Type 23 frigate travelling at a safe and realistic speed.

Table 3 Power/weight for different handling characteristics.

Model Response	Power/Weight Ratio	
	Watts/ Pound	Watts/ Kilogram
Steady	< 1	< 2.2
Handy	1 – 2	2.2 – 4.4
Lively	2 – 3	4.4 – 6.6
Exciting	3 – 5	6.6 – 11

do. This level of performance does demand more care when operating, like anticipating where the model is likely to be in a few seconds. With the 'exciting' ratio of power to weight, the model's speed is impressive without being over the top. It does, however, demand total concentration on what it is doing and any potential problems that might appear in the next few seconds. Not perhaps the best place to start but, like many activities, once mastered, it can be rewarding.

The full-size *Huascar* had a very modest top speed and the model was likely to be in the steady category. Using its estimated weight (5kg or 11lb) a power of 10W would be a good place to start. The Type 23 frigate was less weighty (1.8kg or 4lb), but with more speed expected it would be better in the lively/exciting groups, so the power would likely be 10–20W.

At this stage, with all the numbers looking promising, a project can progress with some confidence. Nevertheless, it is always wise to keep double-checking things. A bit of arithmetic can be a lot cheaper than having to scrap a piece of work.

Chapter Three

Hull Materials, Tools, Adhesives and Joints

These topics are, as most things seem to be, interdependent. A selection in one area can influence what you have to use in another. Different people will have different favoured ways to achieve an objective, but you ought never to think that there is only one way.

MATERIALS

As we are not going down the route of assembling a model from a commercially supplied kit of readymade parts, there is an initial choice to make about the hull of the model. Is it going to be self-built or can a commercial item be used? Many people fight shy of building the hull themselves and buy a readymade hull, usually a plastic or GRP (glass reinforced plastic) product. This certainly speeds the building process up and can overcome possible limitations in available workspace, tools, equipment and time. It does, however, depend on there being a suitable commercial product available. That is, one that matches the size and type of vessel upon which you plan to base your model. To some extent, hull designs can look similar for different ships of the same functional classes. Thus, you can sometimes get away by using a near identical hull. The choice is yours, but using the wrong hull for the type ship you are aiming to create can end up with a model that will never look right.

WOOD WORKS

Wood is a very suitable material for building working models. It has a good strength-to-weight ratio and is easily cut, shaped and can make strong, glued joints. But it is hard to use wood successfully without some idea of its properties. Unlike some materials, it has different characteristics in different directions. Failure to appreciate this can, at best, create problems and, at worst, lead to failures.

A good way to think about wood is that its structure is like a bundle of straws or tubes and they are aligned in what we call the 'grain' of the wood. This explains why liquids readily soak into one surface (the one cut across the tubes and hence has many openings) rather than another. Likewise, it is easier to split wood in the direction that the tubes lie. This is important when a piece of wood is subjected to a bending load; if the wood grain is in the wrong direction, it can easily snap. As a general rule, when cutting parts out of wood, unless told otherwise, have the grain running along the direction of the longest part length.

Wood can be used in different ways, even just carving the whole hull in one piece from a block. This method does leave you with the problem of hollowing it out so that batteries, motor and the radio control equipment can fit inside. This is a lot of work as well as being expensive, since the

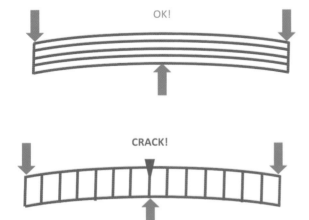

The importance of getting the wood grain the right way to safely accept bending forces.

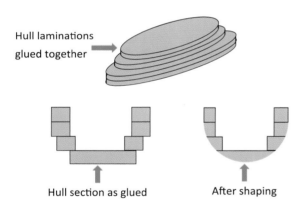

'Bread and butter' hull construction.

at various positions, less the thickness of these strips. This method can reproduce in miniature the shapes of any full-size vessel and, if built with care, give you a light but strong hull. The downside is that it can be a slow process, some people talk of spending an evening fitting a few strips to the frame, especially when the strips need to be individually shaped to accommodate changes in the hull shape.

With SOS models, we are more concerned with the model hull's appearance above the waterline when sailing. Not that this should allow the creation of wildly inaccurate shapes, but allows for a degree of simplification that can sometimes be hard to detect. Luckily, the above-water hull shapes are usually quite simple and, if not perfectly flat sides, have only a limited angling outwards – like the flare often seen in the bow sections – or inwards ('tumblehome'). This allows the hull to be built using wood in sheet form for the sides and bottom, along with some corner reinforcement,

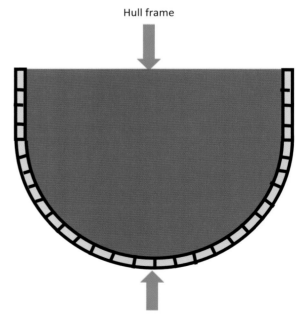

Plank on frame hull construction.

majority of the wood you bought is discarded. This waste can be minimised by laminating the hull from sheets cut to the appropriate shapes but with their centres removed. The sheets are glued together, after which the outer hull shape is formed. This method is often referred to as the 'bread and butter' method.

Even more economical in the use of wood is the 'plank on frame' method of hull building. This involves laying, usually longitudinal but sometimes diagonal, strips over a framework, the framework having sections shaped to match the hull shape

Removable deck section that plugs into hull

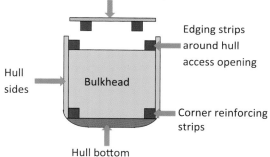

Hull sides

Bulkhead

Edging strips around hull access opening

Corner reinforcing strips

Hull bottom

Sheet balsa hull construction.

which enables a more curvaceous hull section to be made. With care, this simple method can make hulls that closely match the full-size shapes.

WHAT WOOD?

It is unlikely that you can go into a local DIY store and find suitable sheet wood to make a model

boat. Planed square edged (PSE) timber can usually be obtained and could be used in this hobby, but it is usually too thick to allow any of the bending and twisting that we might want to do. It also tends to contain inconvenient knotholes and is not always flat and straight. Sheets of plywood can be found in these stores and again can find some use in this hobby. However, what can be perfectly suitable for domestic construction projects, may be rather rough and less flexible than we might prefer in this hobby.

In many ways, balsawood is ideal for working models. A wide range of accurately cut sheets and strips are available and very strong glued joints are possible. While power tools might be nice to own, most cutting and shaping of balsa can be done with hand tools. Good design and careful construction of a balsa design will produce a durable model. I still have one that is in its fifth decade and still going strong.

Against balsa can be that it is soft and easily damaged with rough handling. In fact, balsa can

HMS Lagos – *over 40 years old and still sailing.*

be obtained in a wide range of densities, but by sticking with medium grades (which can be specified when ordering, here I recommend SLEC in the UK), you can minimise this. Indeed, some of the denser balsa is more akin to lumber from the DIY stores. Finally, when properly sealed and finished, a balsa surface can be surprisingly 'ding resistant'.

If you can select your own balsa sheets, there are a few things to watch out for. A uniform straight grain pattern with no changes in stiffness or density will make cutting out and using balsa sheets and strips much easier. A 'swirling' and variable grain pattern can give you problems.

Sheets and strips also ought to be straight with square edges and no twists or bends obvious. Successfully building models with balsa usually demands these qualities. You can sometimes accommodate a badly shaped sheet or strip, but it needs extra effort and could still leave you with a weakened and distorted structure.

Plywood can be used for these models, but it really has to be birch plywood of the type sold through the hobby trade. This has a smooth surface finish and is available in thicknesses from 0.4mm ($^1/_{64}$in) up. Being made from multiple laminations with their wood grains crossed over, it is tougher and more crack-resistant than plain wood. It is a little harder to cut and shape than balsa and a good alternative can be 'lite ply'. This looks like plywood but has only three laminations and uses a less dense wood than birch. The result

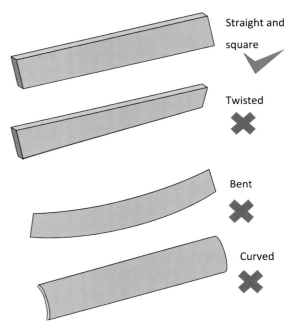

Balsa sheets to avoid.

is something stronger than balsawood but easier to work than plywood. It also has the advantage over balsa that you can get it in larger sheets.

CUTTING AND SHAPING WOOD

The term 'razor blade carpentry' is sometimes used to describe working with balsawood. Please ignore this, as working with razor blades can be dangerous and, in this hobby, totally unnecessary. A good, simple hobby knife should last a lifetime – well, the handle will, but replaceable blades need changing as soon as they fail to cut in a clean and effortless fashion. Plywood and lite ply can also be cut with a hobby knife, but being harder than balsa, greater effort is needed. Something more substantial can be easier and safer to use and I favour those knives with blades that can be replaced or the ends of which can be snapped off to expose a fresh cutting edge.

Accurate, straight cuts need two things, assuming the blade is sharp, of course. That is, a straight

Straight uniform grain pattern – good

Uneven grain pattern – bad

Avoiding poor wood.

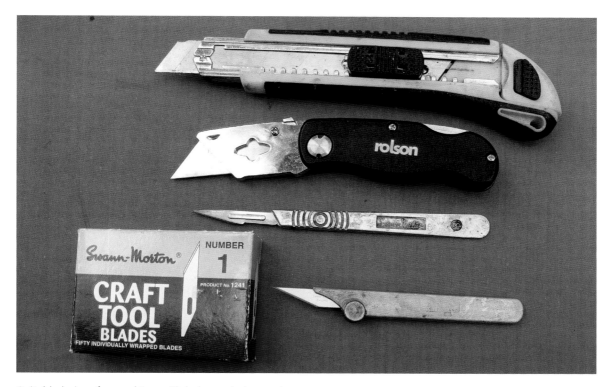

Suitable knives for working with balsa and plywood.

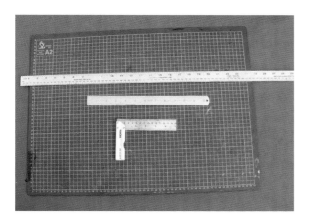

Metal rules, set square and cutting mat.

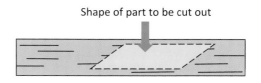

Shape of part to be cut out

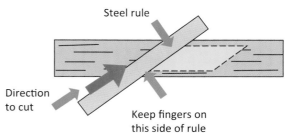

Steel rule

Direction to cut

Keep fingers on this side of rule

The safe way to cut wood.

edge to cut against, and a metal rule is the best way to achieve this. Plastic or wood rules ought not to be used – the blades can cut into them making them no longer straight-edged with the possibility of the blade slipping and causing injury. For safety's sake, your hand ought to be on the opposite side of the rule to the edge you are cutting against. In addition, making accurate cuts at right angles is a lot easier with a metal set square. Unless there is a reason not to do so, the blade should always be perpendicular to the surface being cut through.

The second thing is a suitable firm and flat surface to work on. The term 'kitchen table modelling' has occasionally described modelling in the home without the benefit of a workshop. However, cutting pieces up on the kitchen table, even if you can cope with the domestic displeasure it will likely generate, is not too safe as the knife can be deflected by the table's wood grain. A better idea is to use a 'self-healing cutting mat'. These are made from plastic, and have a surface soft enough to let the tip of the blade cut into but not through it. This prevents damage to the surface you are working on and, as a benefit, makes the blades last longer. In addition, as the name suggests, these shallow surface cuts in the mat heal up to restore a smooth surface.

Wood's preference to spilt along, rather than across, the wood grain can try to deflect a knife blade. For this reason, it is a good idea to try to position the rule and cut the wood so that any tendency to do this is stopped by the rule. It is not always possible, in which case firm but careful control of the knife is needed. I once saw this technique described as being the 'hands of a midwife' – gentle but firm.

Balsa thicknesses up to 6 mm (¼in) can be cut accurately, possibly with multiple strokes, with the use of a metal straight edge and knife. Above this thickness, it can become harder to make square and accurate cuts, especially when cutting strip. A razor saw sold for this hobby is very useful in this situation but a small, fine-toothed hacksaw can be a good substitute. A mitre box can be handy for ensuring consistent square and angled cuts.

SANDING

Even though the surfaces of plywood and balsa sheets and strips should start with a smooth finish, all models built with wood will need some sanding to create the final shapes required and to produce a level surface, especially at joints. This usually starts with a coarse grade of sandpaper to remove the bulk of the material quickly, progressing to a medium grade and finishing off with a fine grade. Packs of mixed grades from a DIY store are perhaps the best way to start.

Using sandpaper held in your hand is a sure-fire way to produce a poor surface finish. Uneven pressure between the paper and the wood, usually caused by finger tips, can create surface depressions and grooves when you want a flat finish. Sanding blocks that have the sandpaper fixed to a flat surface can be used to avoid this problem. It is easy to make up your own blocks from suitable pieces of timber and wrap sandpaper

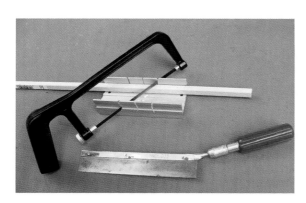

Saws and mitre box for accurate cutting of thicker wood.

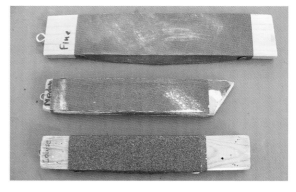

Sanding blocks made by wrapping sandpaper around timber.

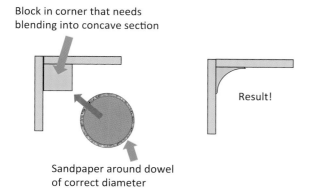

Block in corner that needs blending into concave section

Result!

Sandpaper around dowel of correct diameter

Sanding to produce a curved surface.

around them. Mine have the paper secured with drawing pins, which allows for easy replacement.

Sanding convex, curved surfaces can be done with blocks, but restrained force has to be used to avoid creating 'flats'. Concave shapes clearly cannot be sanded with a flat block, but a neat trick is to wrap sandpaper around a curved shape, a rod or tube of the right diameter and length to handle.

ADHESIVES AND JOINTS

Glues can work by different mechanisms. Solvent adhesives that locally dissolve the surfaces to be joined and then quickly solidify can bond plastics in what is literally a weld. A mechanical joint can be created where the glue penetrates into surface irregularities before hardening to lock the parts together. Using these glues usually demands the surfaces to be clean and grease/oil-free and roughened to create the needed surface irregularities. This is how two-part epoxy-type adhesives work. Glues can also use the fact that atoms and molecules at the surface are not always fully bonded to the ones beneath them. Some glues can latch on any spare surface bonding links to lock themselves in place.

Wood and other porous materials can use a different method. If the glue is applied in a liquid

state, it can be a viscous but nonetheless 'runny' state, it can penetrate into wood before setting into its rigid form. If the joints are close-fitting, as they ought to be, then you have a structure where it changes from wood to a wood/glue matrix and back to wood. A characteristic of such joints is that should they be overstressed; failure will not usually occur at the joint between the parts but in the material nearby, thus explaining a glue manufacturer's sometimes claim of 'stronger that the material it joined'.

WHAT GLUE?

Like many situations, there is no one perfect glue to use on wooden structures. If you have had good results with one type or make of adhesive, then it can be used with confidence. The requirements are to combine a good bond strength and ease (and safety) of working with an acceptable curing time. Economy is also worth considering, but not at the risk of glued joints failing at a later date.

You might be surprised that 'totally waterproof' is not in these requirements. Any model boat should have an exterior surface that water cannot penetrate before it reaches the point of being placed in the water. If not, then no waterproof glue is going to save it from a short and troublesome life.

One type of glue that has proven itself in this type of model is the 'all purpose, weatherproof wood adhesive' sold for general woodworking, joinery, furniture and other domestic uses. It gives you a convenient time to assemble joints and to secure them. While some types may claim to produce joints that can be handled after a short time, leaving them several hours (overnight is a convenient period) is advised. The glue dries clear and joints are strong but easy to sand. The absence of smell and any excess (which usually means accidents) can be wiped away with a damp cloth, and hands can be washed clean, makes it generally acceptable for indoor use.

The term 'weatherproof' might still be a worry, but, in practice, it means that it can be used for external use and getting wet from rain is no problem. However, it cannot be left permanently immersed in water. Now this could only be a problem if exposed glued joints in the model remain in a wet or damp condition all the time. Unless you plan to leave your model permanently afloat, this would only occur if you stored it between sailing sessions without checking for any water inside the hull and letting it dry out.

OTHER MATERIALS

It may be possible to build a model totally out of wood but this would seem to be pursuing a form of material masochism a shade too far. Some metal is going to be used in the model and two-part epoxy adhesives are usually the best to use. The metal surfaces have to be clean and grease/oil-free for the adhesive to create the strongest bond. Abrading the area to be glued with a file or abrasive paper is usually the best way. It also creates surface irregularities into which the epoxy can penetrate to form a mechanical joint. If gluing a metal part into a wooden structure, then one of the slow-setting epoxies can be best. This gives it time to penetrate into the wood pores and creates a stronger bond.

Plastics, in all forms, can be gainfully used in our models. The ideal adhesive would be one of the solvent types: apply to the parts, press together and what is virtually a 'welded' joint is produced. One problem is that there is no universal solvent adhesive for all types of plastic. The best thing to do if unsure is to make a trial joint and test it. The wrong adhesive may hold the parts together but fail as soon as any load is applied. When joining different materials something like a general-purpose contact adhesive will often suffice.

One material that can find a lot of use in this hobby is card. It is relatively economical and free if you make the habit of checking things like

packaging before consigning it to the recycling bin. It might seem to have limited use with its tendency to absorb water and turn into a soft mushy state. However, if used in the right places and with waterproofing, it can be surprisingly durable. As card is porous, it might be tempting to use wood glue on card, but being water-based, it could cause the card to swell and distort. Better adhesives are the solvent-based contact types, which can produce strong joints between card and other materials.

MODEL JOINTS

No matter how good the glue is, it should never be a substitute for a properly made joint in any structure. A joint has to be designed to cope with the forces acting upon it. Significant forces

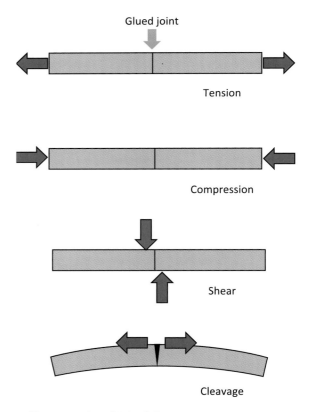

Different modes of joint failure.

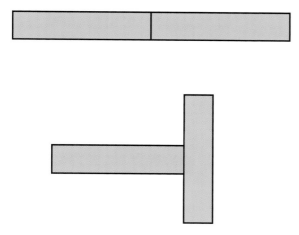

Two types of butt joint.

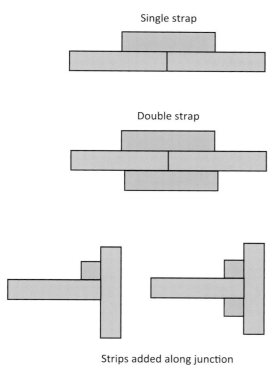

Single strap

Double strap

Strips added along junction

Reinforced butt joints.

can be applied to glued joints during construction, as well as during the model's operation. Joints have to resist four types of failure, as shown the pictures on page 28. In all these

situations, the importance of good-fitting joints is critical and relying on adhesives to fill any gaps should be avoided.

How glued joints are designed can affect their strength. Butt joints are formed when the two pieces are simply glued together. This could be end-to-end or edge-to-edge, but often the two pieces are at right angles to each other. Butt joints can be greatly strengthened by adding strips to one or both sides of the joint. This allows stresses to flow more smoothly across the joint.

SURFACE SEALING

Dropping a bare, wooden hull into the water is an obvious no-no. Apart from stopping water soaking into the wood, sealants are needed to produce a good surface for the final coats of paint and can offer a surprisingly effective toughening effect, not perhaps making a wooden hull impervious to damage, but limiting and localising the effect. In the event of 'accidental' ramming by another model, it can give you enough time to safely recover your model. Yes, it might seem pessimistic but it does happen.

Primers and undercoats used for domestic decorating can be used according to the manufacturer's instructions. I would suggest that the first coats are thinned a little to ensure that they penetrate well into the wood grain before starting to dry. This will ensure a good bond with the wood. Several coats with light sanding between each might be needed to produce a smooth finish and prevent the wood's grain from showing through the final coat of paint. The sight of wood grain in what should be a model of a ship built from metal never looks right!

My schoolboy experiences with balsa and tissue-covered model aircraft has led me to using different wood-finishing techniques. Clear cellulose dope, as used on both model and full-size aircraft, can produce a quick and tough finish on wood surfaces. Again, the initial coats should be thinned for better penetration into the wood.

Purpose-made cellulose thinners can be bought from motor factors quite economically. Light sanding is needed, especially on balsawood, as any previously loose surface fibres become rigid when the dope dries. One problem with cellulose dope is that, being solvent-based, it gives off fumes when drying, making it a job for outdoors or at least a well-ventilated space, definitely not inside the domestic home. However, this solvent does mean that it dries quickly and each coat tends to partially dissolve into previous coats making it a quick job to apply three or four well-bonded coats. One final point is that dope and thinners are inflammable and need storing in a suitable place.

If the solvent fumes or availability prevents the used of cellulose dope, then alternatives are available in the form of clear acrylic varnishes. They have a milky white colour but dry to a clear finish and, being water-based, without any troublesome odour. One tip for using these sealers is to decant some into a glass jar and fix a suitable brush into the screw-on lid such that the bristles are covered with the sealer. This has the dual advantage of avoiding the brushes drying out between coats and the lid can stop drips from running down the handle on to your hands and clothes.

Some surface wood-grain effects might still be visible at this stage and you have two options. A 'sanding sealer' can be used and commercial products are available, but adding something like baby powder or fuller's earth to the sealer is effective. The powder will stiffen and thicken the sealer noticeably and can be thinned to create a better consistency for brushing. Again, sanding between coats of sealer should produce a smooth, grain-free finish.

The second option to achieve a grain-free surface finish is to apply a layer of tissue to the external hull surfaces. The best tissue for this is the lightweight type sold for covering small model aircraft. Unlike the tissue sold for packaging or other uses, this will cover the hull without readily forming creases and has wet strength to avoid tearing during application. Tissue might seem to be low in strength, but when bonded to a wooden hull with a few coats of sealer, it produces a smooth and surprisingly 'ding'-resistant surface. This will not be totally impervious to damage, but will significantly limit it. For the ultimate in damage-resistant hulls, some modellers go for an even tougher hull finish by covering their hulls with lightweight fibreglass cloth.

SEALING THE INSIDES?

Some modellers like to seal the inner surfaces of the hull to prevent any water that may get inside the model from reaching any bare wood and soaking in. This could be achieved if the whole inner and outer surfaces could be covered and cracks, gaps and pinholes are guaranteed to be absent. I have witnessed models where such defects allowed water to become trapped between the otherwise sealed inner and outer hull surfaces, the result being that the permanently wet wood eventually rots, loses its strength and you have a weak area, if not a hole, to repair. My preference is to try to keep the water outside the hull, and, even if the insides look perfectly dry after a sailing season, leave the model opened up to the air off for a couple of days. Up to date, my wooden models look like they will outlast me!

Sealant in jars with the brushes fixed through the lids.

Chapter Four

Motive Power

A model boat can be powered by different methods, but an electric motor is probably the best, at least to start with, if not for most models. Small internal combustion (IC) engines, usually based upon designs used for model aircraft, can create problems with noise and an oily exhaust (unpopular, if not banned, on many public waters). Add

vibration and possible starting problems inside a scale model covered with many detailed items, and IC engines are best left for functional models that are more robust.

A steam engine might seem attractive, especially for models based on suitable prototypes that match their character. A complete steam plant

Steam power – characterful, but not easy.

consisting of the engine, boiler, burner and associated plumbing, can be bought ready to install and run, but at a price. If control of the engine's speed and direction of rotation is needed, then a multi-cylinder engine is essential. I have successfully built and operated a steam-powered model based on the early TBD (torpedo boat destroyer) vessels but found a few problems. Model steam plants turn most of the fuel's energy into heat rather than engine power, and when fitted inside an enclosed hull they need lots of ventilation. Significant preparation and maintenance before, during and after sailing is also demanded. There is no denying the appeal of a steam-powered model and some people find the extra effort in operating such models to be worthwhile, but for ease, economy and reliability, an electric motor takes some beating.

ELECTRIC POWER

Getting the right combination of motors, propellers and batteries needs some care to match all three items together. It is possible for a model to sail with one or more items mismatched with the others, but it will not be a happy situation. Overheated motors, poor speeds and shorter sailing duration than expected are the usual signs of this problem.

PROPELLERS

Propellers are sometimes referred to as 'water screws', which can give the impression that they work like a screw being driven into a piece of wood. If this were so, then holding the model stationary in the water while supplying the motor with power the propeller ought to be unable to rotate. Clearly, this does not happen as it does rotate and you can feel the forward push it applies to the model (and if you are behind the model, maybe the water it throws on to you).

The propeller has two or more blades set at an angle to its axis of rotation. As the propeller rotates, it acts to draw water in one direction and gain speed. This acceleration of the water creates the force that pushes the model in the opposite direction, Newton's famous Laws of Motion in action. Since water is a fluid rather than a solid medium, this works even when the propeller is rotating but not moving forwards, unlike the wood screw.

The angle of the blades with the axis of rotation is referred to as the propeller's pitch. Setting the blades at a small angle produces a fine pitch,

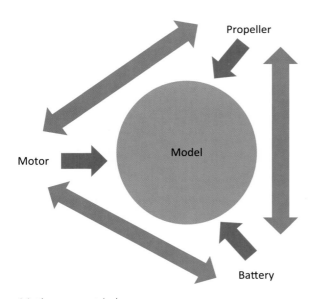

Motive power triad.

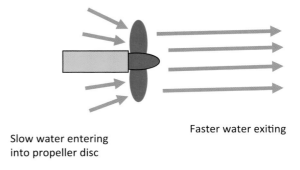

Slow water entering
into propeller disc

Faster water exiting

Water accelerated rearwards by propeller.

whereas a large angle gives you a coarse-pitched propeller. The propeller, and hence the model, would advance the length of the pitch in one revolution in a perfect world.

However, if the advance of the propeller perfectly matched the model's speed through the water, it would, in effect, be freewheeling and produce no thrust. The propeller has to turn faster in order to generate thrust and this mismatch is called the slip and is often quoted as a percentage of the pitch. A figure of 30 per cent slip means that the model only moves 70 per cent of the distance suggested by the propeller's pitch value. As mentioned earlier, even with a stationary model and 100 per cent slip, the propeller can generate thrust; otherwise it could not accelerate from rest.

Propeller Types

When picking a propeller for a new model, you might be able to use the experience gained from similar previous models. However, if it is something outside this comfort zone, then you have to make a guess – hopefully a good guess. One difficulty can be that many propellers are advertised with only their diameter and number of blades being quoted. Luckily, a reasonable pitch/diameter ratio can be guessed as 1.0. Finer pitched propellers will have a value less than this, coarse pitched having a greater one.

For general guidance, and assuming you plan to build neither a very small nor an excessively large model, a propeller diameter in the range of 25–40mm (1–1.5in) might be a good starting point. This could seem a rather small size range, but modest changes in diameter can result in surprisingly large effects. This can be shown in the following analogy of two geometrically similar propellers, that is with the same pitch/diameter ratios, but one having twice the diameter of the other. If they worked perfectly with no slip, then both would push a cylinder of water rearwards as they rotate.

In one revolution, this cylinder would have a diameter equal to the propeller and a length that matches the propeller's pitch. It might seem that the bigger propeller, having twice the diameter, would push twice the volume of water rearwards in each revolution. In fact, twice the diameter quadruples the cylinder's cross-sectional area and

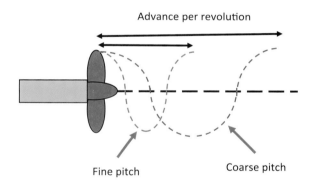

Difference between fine- and coarse-pitch propellers.

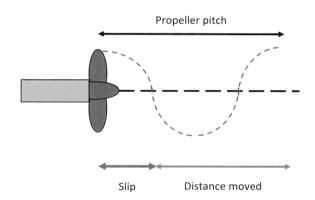

Slip creates thrust.

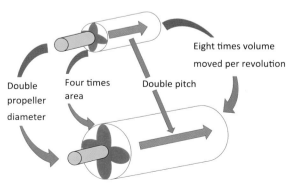

Dramatic effect of increasing propeller size.

(because the pitch is twice a large) doubles the cylinder length, so its volume is eight times that of the smaller one. This is a cubic relationship, so in this relationship, to double a cylinder's volume, you only need to increase the propeller's diameter by about 25 per cent.

As for the number of blades on the propeller, they can vary from two upwards. From full-size practice, it might seem that more blades are better and more efficient. However, our model boat propellers operate at quite high speed with little immersion in the water. In this situation,

Typical 2–3–blade model propellers.

well-designed, two- or three-bladed model propellers work effectively.

Propellers can be made from either plastic or metal, usually cast brass. The latter look very nice when polished and the model is being displayed out of the water, but are much more expensive than the plastic ones and do not guarantee any better sailing performance. As you may have to experiment with a few different types of propeller before getting the right combination of speed, duration and handling, several plastic ones can be bought for the price of one brass one – your choice. As we are working in the SOS zone, it has always seemed better to use a propeller based upon its performance rather than appearance.

A model of a portly vessel that has a modest top speed, such as a tug, might benefit from a large diameter, fine-pitched, multi-bladed propeller. This would allow the motor to drive a large volume of water rearwards but at an efficient low speed. Higher speeds are usually expected with warship models, so they can favour smaller diameter, higher pitched propellers. These can drive a similar volume of water rearwards but at the higher speed that the model needs. This is illustrated with the *Huascar* and Type 23 models. The *Huascar* only needs a modest speed (no more than 61cm/s or 2ft/s; *see* Chapter 2) and a 50mm (2in) diameter propeller rotating at a modest speed would be a good match. The Type 23 frigate model on the other hand had to be capable of reaching over twice this speed (1.3m/s or 4.2ft/s) and with its sleeker hull shape, a smaller 35mm (1.4in) two-bladed but higher pitch propeller was just right.

Different model requirements – different propellers.

BASIC ELECTRICITY

Before moving on to the motor and battery, a few comments on electrical terms might be welcome. This subject is often misunderstood and the incorrect application can result in dramatic and expensive accidents.

- Voltage, measured in volts (V), can be equated to the 'pressure' that wants to drive an electrical charge around a circuit.
- Current, measured in amps (A), is how fast the electrical charge is flowing in the circuit.
- Resistance, measured in ohms (Ω), tells you how difficult it is for the voltage to drive a current through a circuit.
- Power, measured in watts (W), is the amount of energy being transferred per second.

They are related by simple equations:

Voltage = Current × Resistance (or in symbols
$$V = I \times R)$$

If you want to drive more current through a circuit, then increase the voltage and/or reduce the resistance.

Power = Voltage × Current (in symbols $P = V \times I$).

More power calls for increasing the voltage and/or current.

If you need to measure them, then a voltmeter has to go across (parallel) whatever device you are testing. The current is measured with an ammeter placed inline (series) with the item. A small multimeter capable of measuring both voltages and currents is a handy thing to have at both the workbench and pond side. It can save you a lot of heartache and wasted effort, even money!

ELECTRIC MOTORS

Having said how suitable electric motors are for our models, and they can be considered almost 'foolproof' (determined fools will always succeed in breaking anything), it is surprising that many people understand little about how they work. They can come in many shapes, sizes and types, but all depend on electromagnetism, that is, when an electrical current flows, a magnetic effect is also produced. The simplest demonstration of this is to wrap some insulated wire around an iron bar. When the ends of the wire are connected to a battery, the bar becomes a magnet with a north pole at one end and a south pole at the other end. When the electrical circuit is broken, this

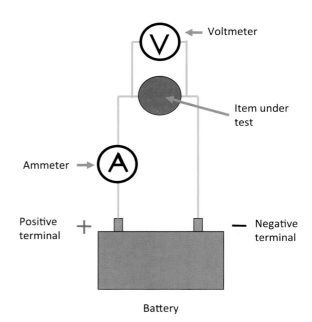

Basic electrical circuit.

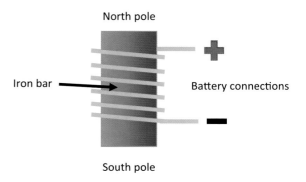

Electromagnet – the heart of a motor.

magnetism disappears. If the battery connections are reversed, the current now flows through the wire in the opposite direction, then magnetism is restored but with the poles also reversed.

If the iron bar were pivoted at its centre and placed between the poles of a permanent magnet,

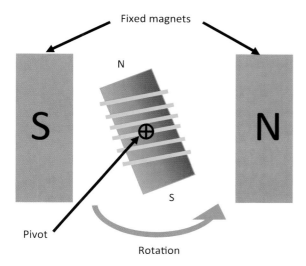

Electromagnet becoming a motor.

then as soon as current flowed the poles created by electromagnetism would cause it to rotate. This would not be a motor yet, as movement would stop as soon as the poles were aligned north–south and south–north. Keeping it rotating requires the current to be repeatedly reversed at just the right moment in the rotation. This leads to two different types of motor available for our models: brushed motors and brushless motors.

Brushed motors achieve the switching of the current by having the coil windings connected to something called a split-ring commutator, which has separate segments fitted to the rotating motor shaft. Two fixed brushes carry current on to the commutator and as it rotates, it switches the current at just the right position to keep the motor working. In practice to ensure these motors will always be in a position to self-start and make them smoother running, they have more than two poles. Three or five poles are the commonest types of motor and an opened-up motor is shown in the photo below (not a practice recommended to be done lightly, as it is not so easy to reassemble).

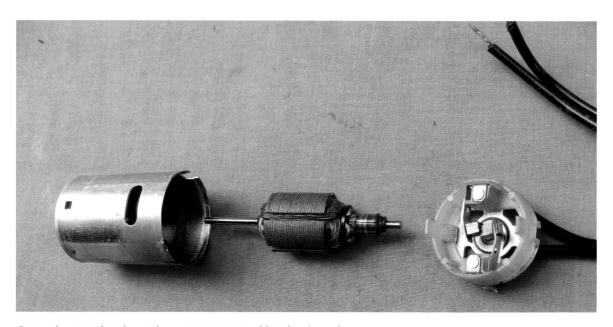

Opened motor showing poles, commutator and brushes in end cap.

The brushes have to rub firmly on the commutator to ensure that the current flows without interruption. This creates extra friction and wear, which can be mitigated by good design and things like carbon brushes. However, brushes can be eliminated altogether with the design of brushless motors. The switching action of the commutator is replaced with electronic switching. This requires the use of dedicated control units, which must match the motor; brushed motors just need a voltage applied to their terminals. For high powers, brushless motors are superior, but there is no clear advantage over brushed motors in many types of scale models.

ELECTRIC MOTOR PERFORMANCE

Despite the apparent simplicity of electric motors, they are easy to install and operate badly through a lack of basic knowledge of their characteristics. This is not a difficult subject if you avoid becoming too technical. The three diagrams on the right ought to explain how to ensure your models run well.

The first thing to realise is that when initially connected to a battery, it will accelerate up to some maximum speed and the current drawn falls to a minimum value. Quite naturally, these are called the no-load speed and current. As an increasing load is applied, the motor slows and the current rises. It is a steady change – more load leading to lower speeds and greater currents. However, at the same time the torque (the twisting effort) produced by the motor increases. This carries on until the motor stops rotating and is said to be stalled. The current is a maximum at this point (stall current). This is not a good place to be, as the wires inside the motor will be getting hot, very hot and distressed, possibly to the point of ruining the motor.

The actual mechanical power output of the motor is not a constant. When running at the no-load speed, the motor is doing no external work and all the electrical energy supplied is being

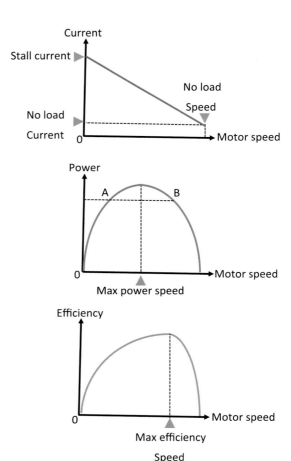

Why high speeds are good for motors.

absorbed by internal losses like friction and electrical resistance. It is the same when the motor is stalled; it is doing no work but this time the massive current is just heating things up. Clearly, it must produce some power output to justify its existence between these two extremes. In fact, the power it can supply to do work, like moving your model around, will increase as the motor is slowed down from the no-load speed. It will peak at something like half of the maximum (no load) speed and then fall to zero when stalled.

If you needed the maximum power that the motor is capable of producing, then it has to run at this maximum power speed. However, this will usually be way over the needs of most scale models. If you use less than this maximum power, then it can be seen that you have two possible motor speeds to supply this power: points 'A' and 'B' on the power graph. It seems sensible to settle for the most efficient situation given the choice of two ways to achieve the desired power.

In this case, efficiency is the ratio of mechanical power output divided by the electrical power input. Both points 'A' and 'B' have the same power output, but looking at first graph shows point 'B' to have a much lower current and hence power input (W = V × I) and so is much more efficient. This explains the skewed shape of the efficiency against motor speed graph. Running the motor at about 70–80 per cent of the no-load speed is a good place to be, high efficiency producing a cool motor and long battery life. It is also safer, since if the propeller were to become fouled with something like weeds, the motor would slow but actually produce more power and torque, and might free the obstruction or at least allow recovery of the model. Picking up weeds while running it at the maximum power speed will slow the motor into a lower power and a less efficient condition, and maybe even stall it.

MOTORS

The most widely used brushed motors are the can-types, that is, a cylindrical metal body with the drive shaft protruding out of one end and the two electrical connections at the other end. They first

Typical can-type motors.

came into the hobby from the Japanese Mabuchi manufacturer, although now many other manufacturers produce similar types but sellers often fail to distinguish differences. This can result in confusion, as one motor may be a perfect match for your model, while other apparently identical motors can be screaming overpowered beasts or weak efforts.

There are two motors in the Mabuchi range that make good starting points: the RE-540 and the RE-385. The former is a three-pole motor with a 36mm (1¹/₂in) diameter body, with a length of 50mm (2in). On 6V, it has a free-running speed of around 7,500rpm and when loaded to draw currents up to 2A ought to give good efficient service. The smaller RE-385 (28mm diameter, 38mm length) has five poles, runs at 5,500rpm on 6V and currents around 1A are suitable.

There is another type of brushed motor that can be very useful, these are rock crawler-types. They were intended for RC cars used to traverse across very rough terrain and required a motor with modest but controllable speeds, along with more than adequate torque. This is achieved by winding more turns of finer wire around the poles of the motor, creating a more powerful magnetic effect at lower currents and motor speeds. This can be ideal for some model boats where they can directly drive propellers that would otherwise be too large.

These rock crawler motors can have a sealed can-type of body, but others may have an exposed brush gear at one end. This usually means that replaceable carbon brushes are used, which can mean greater efficiency and longer motor life.

BATTERY

The electrical energy to power the motor and thus drive our models along comes from a cell. The chemicals within the cell want to react, but this can only occur when charge flows between the positive and negative terminals of the cell. The cells generate a characteristic voltage between the terminals, dependent on the chemical reaction. In many cases, the individual cell's voltage is too low to supply the desired electrical current and so the cells are connected in series. This, in effect, increases the 'pressure' to drive the current through the motor, which means more speed and power. For example, the lead-acid cell (widely used in automobile batteries) has a voltage of around 2V. Connecting them in series (positive terminal of one cell to the negative terminal of the next cell and so on) can produce a battery with an output voltage of any multiple of 2V, such as 6 or 12V.

The cells can be divided into two types, primary and secondary, the difference based on the chemical reaction. In the primary cells, the reaction will only progress in one direction. When the chemicals inside the cells are consumed, it stops working and is often termed to be 'dead'. Primary cells are often referred to as 'dry' cells; this is due to their sealed nature compared with the early secondary cells that contained liquids. A common type is based on alkaline chemistry and has a cell voltage of 1.5V, thus it can make up batteries

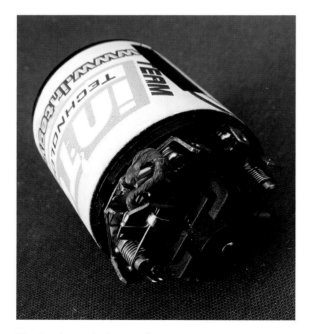

The 'rock crawler' type of motor.

with voltages of any multiple of this value. They do have a limited ability to supply a steady current. This usually makes them a poor choice for motors that drive the model, but can be perfectly suitable for a transmitter or receiver power supply. They also have the useful characteristic of the battery voltage falling steadily with use, which can be used as a warning before they become dead and need changing. Many transmitters have a built-in warning system to tell you when this is about to happen before control is lost. If you use these dry cells, then taking a spare set when sailing is not a bad idea.

Secondary cells have a different reaction that can be reversed. When delivering a current (discharging) they act like a primary cell, but when the chemicals are consumed (the cell is usually termed as being flat) it is possible to recharge it by driving current in the opposite direction.

As well as a characteristic cell voltage, they also have a property termed capacity, which is very important. This governs the time for which a cell can deliver a current before becoming flat. A useful way to think about cell capacity is with two identical sized buckets, one with a small hole in the bottom and the other with a larger hole. If both start full of water, then you can expect the bucket with the larger hole to empty first because the flow rate of water through the hole will be greater. This leads to the idea of the relationship between capacity and flow rate:

capacity = flow rate × time to empty

Hence, for any given capacity, doubling the flow rate will halve the time to empty. Conversely, keeping the flow rate constant and increasing the bucket's capacity will allow water to flow for longer.

Transferring this idea to a cell, we have to use amps as our flow rate of electrical charge and hours as the time to empty. Therefore, a cell's capacity is measured in amp-hours (often abbreviated to Ah). Thus a 6Ah cell ideally can deliver a current of 6A for 1h, 3A for 2h or any

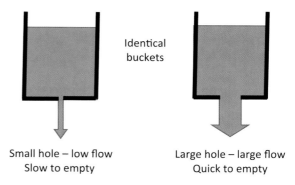

Small hole – low flow
Slow to empty

Large hole – large flow
Quick to empty

Bucket analogy for battery capacity.

combination where the product of current (A) and time (h) equals 6. I do say 'ideally' as a cell's effective capacity can be noticeably reduced if it is rapidly discharged.

Lead-acid batteries are a popular source of power for ship models, with 6 and 12V being the most common. The acid is usually in the form of a jelly or paste inside a totally sealed case; no liquid acid to spill and, therefore, much safer and easier to handle. The cases are usually in a rectangular shape and, being sealed, this means they can be placed in any convenient position, often flat on one side to keep a model's centre of gravity low. The power is supplied from two terminals, which usually match the small spade connectors that are often used in automotive and domestic appliance applications.

Nickel-based batteries (originally nickel-cadmium but now nickel metal hydride chemistry) are a popular choice. The individual cells are usually in a cylindrical shape and can directly replace primary (dry) cells in some applications, such as powering the radio control system. They do have a slightly lower voltage, 1.2V, but in this application, it is not usually important. Battery packs are made up from cells welded together and the six-cell pack, which will produce 7.2V, is a popular size. Wide ranges of battery capacities are available up to several amp-hours. One point worth mentioning is that some of these cell capacities are quoted in 'milliamp-hours'. This is no more than using

A range of lead-acid batteries.

milliamps (one-thousandth of an amp) and so is just the amp-hour capacity multiplied by 1,000. In other words, 3,000mAh is the same capacity as 3Ah.

Recharging Batteries

Batteries made from secondary cells can be recharged easily and potentially give many hundreds of discharge/recharge cycles with our models. The initial extra expense compared with primary cells (which probably could not produce the desired model performance and duration anyway) can be recovered by the repeated use of these batteries.

Lead-acid batteries ought only to be recharged at a modest current to avoid damaging the cells.

The best recharging current is the '10h rate', that is one that would, in theory, fully recharge a totally flat battery in 10h. This is a simple enough calculation; for example, using a 6Ah battery:

recharging current = battery capacity divided by 10

$$\frac{6}{10} = 0.6A \text{ (or 600mA)}$$

As fully discharging a lead-acid battery is not recommended, since it could cause chemical damage inside the cells, you might be left with the problem of having to recharge a battery that is only partially discharged. Not to worry – these sealed batteries should be designed so that they

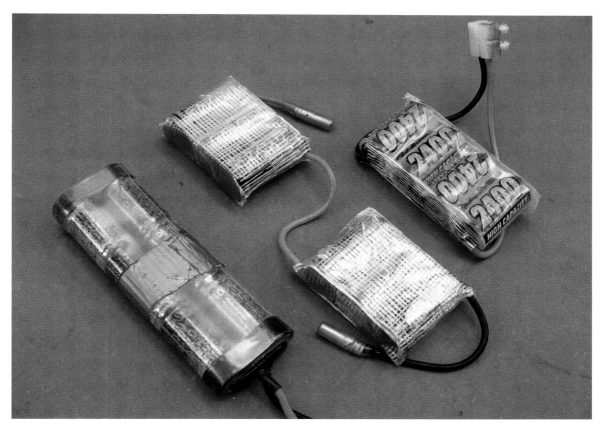

Battery packs made up from different combinations of NiMH cells.

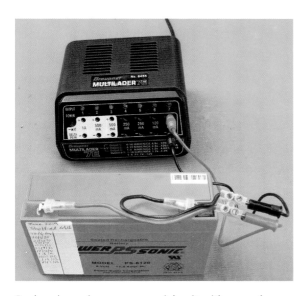

Recharging using a commercial unit with several different output currents.

can withstand overcharging at this 10h-rate. In fact, to ensure that batteries are fully charged (there are always some losses somewhere in the system), you are recommended to recharge at the 10h-rate for 12–14h.

It is important to recharge lead-acid batteries promptly after use as they suffer from an effect called self-discharge. Even just sitting there not apparently doing anything, a small current will be leaking through the battery. If left for a few months, even after being fully recharged, I give them a short top-up recharge, maybe the 10h-rate for 3–4h. It is not a bad idea to stick a label on the battery case and write down the date of the last recharge.

Nickel batteries can also be charged at the 10h-rate safely but also at much faster rates. This requires special battery chargers that monitor

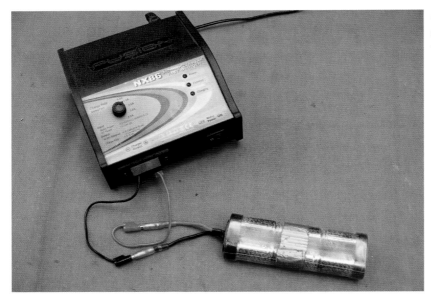

Fast recharging nickel battery packs with an automatic, and safe, charger.

the voltage of the battery pack and can detect when the cells are fully charged and then switch to a lower, safe current. This process is referred to as delta peak charging. Using this method, it is possible to safely recharge these batteries in an hour or even less. Attempting such fast recharging without using such a charger, in other words guessing the amount of charge needed and hoping to switch off at the right time, is tempting fate.

Lithium-based batteries have entered the hobby world and can offer light weight and high-power levels. They have been widely taken up by the model aircraft and car fraternities where these qualities are highly desirable. Operators of very high-speed model boats use them, but they are not essential in many scale-types of boat. Recharging of these batteries needs a dedicated type of charging unit, which monitors individual cell voltages. It is also vital that they are never over-discharged or they can be ruined. As most scale models lead a more sedate life, and usually need extra ballast to float stably on their water-line anyway, the weight of lead- or nickel-based batteries doesn't go to waste.

COMBINING ITEMS

Most warships have quite sleek and slippery hull shapes, as befits their functions. Models based upon them can also be quite easy to drive at a realistic speed. In Chapter 2, how to estimate a model's speed to produce a realistic wave pattern was described. This is usually a comfortable and easily reached speed. The relationship between a model's speed and the power needed is approximately a cubic one. Therefore, doubling the power might only increase a model's speed by about 25 per cent. However, doubling the speed could require 'eight' times the power. In fact, drastic overpowering could lead to models trying to climb out of the water with the bows rising and stern sinking. If not dangerous and unstable, it looks very silly.

I will admit to taking a few different propellers with me on the first test sailings of a new model, as it is impossible to get it right every time. Luckily, if the propellers are part of the same family (that is they have the same pitch/diameter ratios) then a small change in diameter can create a noticeable change in the model's performance.

Chapter Five

Radio Control

A range of transmitters, different makes but all operate in similar fashion.

You can buy a radio control outfit, plug the parts together, switch on and have it operate flawlessly without knowing any more than that. This is fine, and let us be honest, it is the way many people function these days, but if it fails to work they are usually left with the insane approach of repeatedly switching on and off in the hope of a different outcome. Like most things, a little understanding (which does not have to be highly technical) can often solve many problems.

A LONG HISTORY

The ability to use radio control to control a model boat was first demonstrated by Nikola Tesla in New York way back in 1898. It remained something of a novelty and the province of dedicated technical experts until the mid-twentieth century. One effect of World War II was the rapid development of radio/radar and people trained in this technology. By the 1950s this had fed into the model-making world and commercial radio-control equipment became readily available.

The early equipment was very limited, operating on a frequency of 27MHz (27 million cycles per second) and only one model could be operated at a time. Control was very basic at first: a single push-button on the transmitter could operate a single control, usually the rudder in a model boat. Even this was usually limited to cycling between full-left–neutral–full-right, although with ingenious electromechanical devices, it was possible to select right or left rudder and even control the speed and direction of an electric motor. This required the operator to press the transmitter button in the correct sequence, any mistakes resulting in unwanted but possibly amusing manoeuvres.

Things improved with transmitters starting to feature switches that could reliably operate the desired rudder movements and motor control.

Even better was the ability to have precise frequency control by using a matched pair of crystals: one in the transmitter and the other in the receiver. At first, this allowed up to six models to be operated at once, then, with improved electronics, up to twelve models simultaneously.

As radio-controlled models became more numerous, even twelve possible frequencies was a serious limitation. This resulted in two further frequency bands being made available. For safety, model aircraft were given the exclusive use of 35MHz, while surface models (boats and cars) would use the 40MHz band. This meant that along with the 27MHz band, up to 42 model boats could sail simultaneously – provided that they checked that no frequencies were duplicated. Things are even better now as we now have radio-controlled (RC) gear available on the much higher 2.4GHz frequency and each transmitter should only communicate with its own dedicated receiver. With this equipment, there is no need to check frequencies anymore.

SMOOTH PROPORTIONAL CONTROL

It is not hard to see that steering a model boat with nothing between neutral and full-rudder positions would lead to a poor response when sailing. With a slow model, this might not be too obvious and 'blipping' the rudder command could produce gentle changes in course. However, anything faster was bound to look 'jerky' and make smooth, controlled sailing difficult. Luckily for us model boaters, this problem was even more serious for people trying to fly RC aircraft and a solution was found.

FEEDBACK PROPORTIONAL CONTROL

The handheld transmitter has two sticks on the front of the case: one moves horizontally for rudder control and the other vertically for motor control. When the system is switched on, the servos might momentarily twitch but then stop moving. As soon as a transmitter stick is moved, then the output arm or disc on one servo will also move. A small stick movement produces a small rotation of the output; move the stick further and the output rotates more. Push the stick the other way and the servo rotates in the opposite direction. You can imagine that these transmitter sticks are directly connected to the rudder and throttle in a model, as the servos perfectly and virtually instantaneously mimic the stick movements. The two sticks are usually spring-centred back to the neutral position. This is ideal for the rudder, as it should return the model to running straight ahead when your thumb is not supplying any sideways' force. The throttle stick could be set up the same way, centre being 'off', and any ahead or astern power requiring a definite push on the stick. Some people prefer the throttle stick to stay where it is placed, and usually the centring springs can be disconnected and a ratchet installed. This does, however, involve going inside the transmitter case.

The electronics inside the transmitter case generate a continuous sequence of radio pulses. The length of these pulses is governed by the position of the sticks, as they are mounted on potentiometers that vary the voltage signal sent to transmitter electronics. These radio pulses are detected by the receiver and, the clever part, they are sorted so that the pulses generated by the position of the transmitter's rudder stick only go to the rudder servo and that from the throttle stick to the other servo. This is why servos have three wires connecting them to the receiver: one for the positive connection to the power supply, one for the negative and the third for this position signal.

The servo uses the principle of negative-feedback to mimic the transmitter stick movement. The servo output is driven by a geared electric motor. The output shaft is also connected to a potentiometer (like the transmitter sticks) and also

A basic two-function RC outfit.

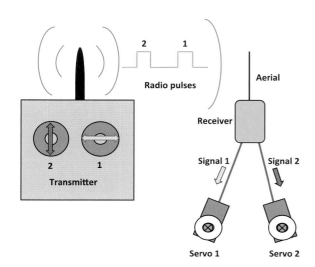

Operation of a two-function outfit.

generates a voltage signal. The radio pulses that the receiver detects have also been converted to a voltage and the servo electronics compares these two voltages. If there is no difference then the servo does nothing. If a difference is detected, it drives its motor to reduce the signal difference from the potentiometer on the output shaft to zero. Thus, the servos try at all times to copy the position of the transmitter controls.

TRANSMITTERS

Virtually all the RC gear used in boats was developed primarily for model aircraft or car use and as a result ought to be robust and reliable. It can have a few features that are not obviously of use in a model boat. Only a two-channel or function RC system has been discussed so far. In practice, most available RC systems can control four or more servos at once. The extra servos can be controlled by switches or rotating knobs/levers on the transmitter case, but often dual-axis sticks are fitted.

In these units, the stick is free to move in the vertical or horizontal directions at the same time. There are centring springs installed to return the sticks to the middle or neutral position, although you can usually disconnect them and fit a ratchet. This is clearly related to their use in model aircraft where the dual-axis movement mimics the joystick in a full-size aircraft. You can ignore one axis on each stick and just use them to control the rudder and motor controls, as in a two-channel transmitter. They can be used for other functions but require good thumb coordination.

Sometimes, as installed in the model, the rudder might not be accurately centred. With the transmitter's rudder stick in the central position, the model might run with a slight turn one way or the other. To make small, fixed adjustments to correct this, the transmitters have trims by each stick. They are small levers or rocker switches that can adjust the neutral value of the transmitter's pulses – very handy for small adjustments but not a substitute for correct installation of the RC system.

Transmitter with two dual-axis sticks.

Transmitter trim switches numbered to match receiver sockets.

In the early days of proportional RC, you had to make sure that the rudder servo was correctly linked to the rudder. It was easy to launch a model on its maiden run and find that the model turned left when you gave it the right-rudder command. Correcting this either required buying a servo with opposite rotation or tinkering with the servo/rudder linkage. Nowadays, transmitters usually feature a very useful servo-reverse switch for their main functions.

There is a different transmitter that is widely used by the operators of RC cars, the steer-wheel type. A wheel that is, again, spring-loaded replaces the rudder stick. Some people find this a more natural way to control a model and it does

have one advantage when you are facing a model travelling towards yourself. It can confuse many, at least at first, that in this situation, if you want the model to turn to your left, you have to push the transmitter stick to the right. With a steer-wheel in the same situation, you rotate it clockwise, which can seem more natural.

For comfortable operation, these steer-wheel transmitters feature a pistol-grip body, with the left hand holding the grip and the right hand the wheel. The throttle control is a trigger operated by a finger on the left hand. This is also spring-centred and needs either a push forwards or pull backwards. These transmitters also feature servo-reversing, trim adjustments.

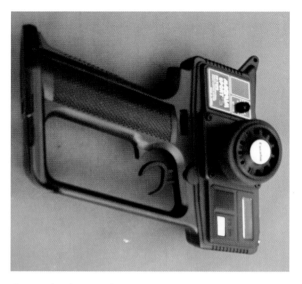

Steer-wheel transmitter.

RECEIVERS

Receivers used to be quite chunky items containing many discrete components, but, due to things like integrated circuits, they are much smaller now. Their usual layout is to have a row of sockets at one end for the servos to plug into.

When the receiver has several servo sockets, it may not be clear which is operated by the different transmitter controls. The sockets ought to be numbered and the transmitter might have matching numbers on its case, for exapmle No.2 could be the horizontal stick movement to control the rudder servo. If it does not then it would be a good idea to put the appropriate receiver socket number on the transmitter case. This makes life much easier when installing equipment into a model.

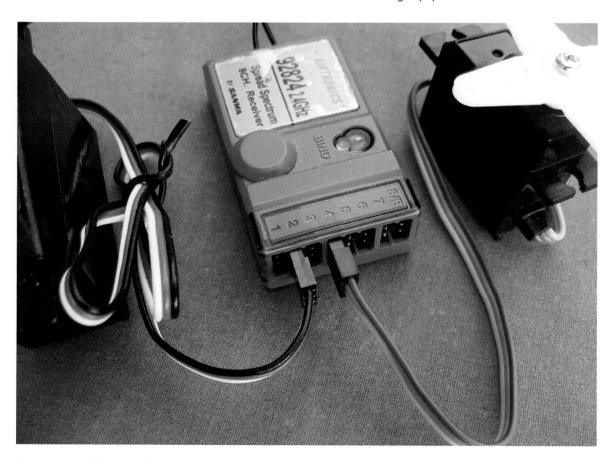

Servos plugged into receiver sockets.

Transmitter sticks and switches numbered to match receiver sockets.

Usually at the other end of the receiver case is the aerial wire. With the 27/40MHz RC systems, this was a long, flexible wire. This could be a problem, as it really needed not to be coiled up inside smaller models and ideally have some of its length in the vertical plane for good reception of the radio signal. Luckily, the modest operating distances for most model boats allowed less than perfect aerial installations to work adequately.

The higher frequencies used by 2.4GHz RC systems allow a much shorter aerial to be used. In fact, it is often only the bared end of a length of coaxial cable that does the receiving. The receivers can have single or double aerial wires; both work well, provided they are not installed below the waterline of the model. This radio frequency cannot penetrate through water as well as 27/40MHz does. Likewise, it can be blocked by conducting materials, such as metals and carbon fibres. The idea of having two aerial wires is to mount them in the model at right angles to each other so that, in the event of one being badly oriented to receive the signal, the other one ought to be suitable. An easy way to achieve this, and to allow quick installation and removal, is to tape the two aerial ends to a piece of card or plastic.

SERVOS

A bewildering range of servos is available for our models. They mostly follow the same form of a

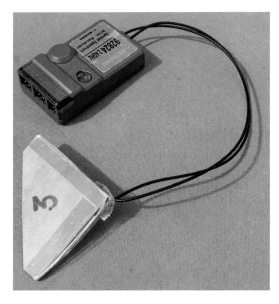

Two aerial wires mounted on card at right angles.

rectangular plastic body with mounting lugs at each end and the rotating output shaft on the top. The three-wire lead comes out of the body and ends in a plug, which matches the receiver sockets. Unless you have space limitations or plan to operate a high-speed model, then the 'standard' size of servo ought to be adequate.

This plug is polarised so that it can only fit one way into the socket to ensure that the positive, negative and signal wires are correctly connected. The small size of these plugs, especially when working inside a model, can make it too easy to fit them the wrong way around. Brute force has sometimes pushed them the wrong way into the sockets and, at best, the servo fails to work, at worst, the receiver is ruined. If there is any resistance to the plug entering the socket, then stop and look closely at the plug and socket.

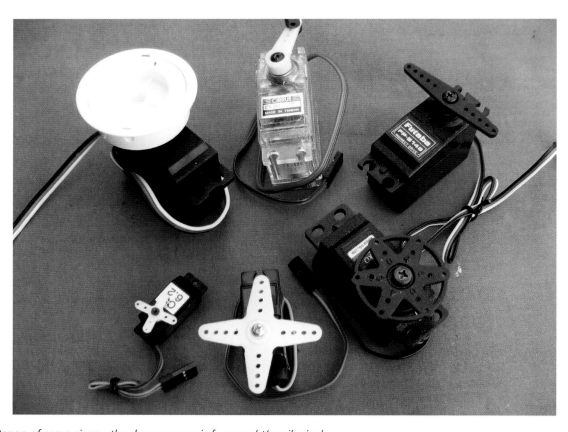

Range of servo sizes – the drum on one is for a yacht's sail winch.

The servo delivers its operating force via a disc or arm, which has holes at varying distances from the axis of rotation. By using pushrods fitted into these holes, the movement and force transmitted can be altered. The further the hole is from the axis, the more movement can be created but the force will decrease. In practice, the rudder forces on most scale models should be modest and well within the ability of standard servos to operate. A good test for any servo–rudder linkage is that, even in the water with the motor running at full speed, it moves with little, if any, perceptible delay to the position commanded by the transmitter. Any sluggishness in the servo movement could be a sign of problems that might lead to difficulties when in an awkward situation.

ELECTRIC MOTOR CONTROL

The simplest form of motor control was just to have a servo operate an on–off switch. This was very limited, although capable of being modified to something more useful with the addition of another switch to give forwards–stop–reverse motor control. This might be suitable for models based on sedate types of vessels, but warship models really needed something better. To be fair, this simple switch control can still be a satisfactory way to control auxiliary functions, such as bow-thrusters, where fine control is not always needed.

Servo-operated, variable resistance controllers allowed for more speed options but wasted some of the battery power by turning it into heat. It was also tricky to sometimes match the speed controller to the motor, propeller and model. Moving the controller from the stop to the lowest speed setting would not always start the motor rotating and you had to 'blip' it with a higher power before dropping back to the lowest. This was never ideal when attempting a tricky manoeuvre. However, it was better than the limited control with just switches and was easy to install and maintain.

Infinitely variable speed control came with electronic speed controllers (ESC), which could just be plugged into a receiver socket. Connected to the model's battery and motor (correctly, as any mistakes are usually fatal), they could run the motor from slow to full speed in both directions. This is usually achieved by pulsing the power to the motor so that it is possible to start the motor turning at dead-slow speeds. Many ESCs can be

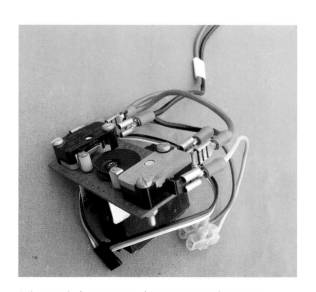

Micro-switches mounted on servo to give motor control.

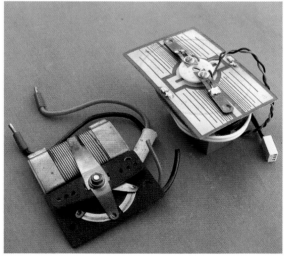

Variable resistance servo-operated motor control units.

used as soon as they are plugged in, but others offer the ability to adjust the position of the neutral (motor stopped) position of the transmitter control and the top speed of the motor.

There is the need to ensure that the ESC has a suitable rating for your model installation. This is the maximum voltage and current it can cope with. The model's battery voltage should not exceed the ESC rating and its maximum current should be well below what the motor is likely to draw, giving some reserve to allow for it becoming stalled by weeds and such like.

One very handy feature of ESCs is that they usually incorporate a battery eliminator circuit (BEC). This takes the model's drive battery and supplies power to the receiver and servos, thus removing the need for a separate battery pack. Even if the voltage of the model's battery is too high to be safely used directly by receiver, the BEC reduces it to an acceptable level. This saves weight and space, along with simplifying wiring, as long as you never let the drive battery go flat whist

sailing. In addition, even with the BEC switched off, some current might still be drawn from the battery, so, when not sailing, the battery ought to be disconnected from the ESC.

EXTRA FUNCTIONS

While getting the rudder and motor controls operating reliably is the first priority, extra functions can be added. This can make use of all those knobs, switches and levers that would otherwise be unused on the transmitter. This can be for lights, sound effects and gunfire (simulated not real!). A simple servo-operated switch could be used, but units where, via some electronics, the receiver's output directly switches things on or off are available. They will have a maximum voltage and current rating that needs to be adequate for your installation.

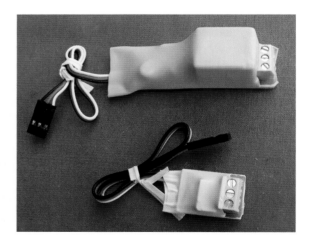

Electronic speed controllers that plug into the receiver.

Switches that plug into the receiver.

Chapter Six

Designing the Hull

Some sort of plan is needed before starting to build the hull of your model. One of the joys of building models with the SOS approach is that you do not have to draw up highly detailed plans to start with. Nevertheless, a drawing of the basic hull in side and plan views is a good starting point.

These drawings need not be full size and something around one-quarter model size is adequate. To keep things neat and square, it can be useful to draw them on graph paper. Drawings, and even suitable photographs in reference sources, can provide the dimensions, proportions and shapes for these plans, but are inevitably not a convenient size to use directly.

SCALE RULES

My early attempts at designing and building working model warships made use of some plans intended for building miniature (non-working) models at 1/1,200 scale. They were side and plan views of vessels but clearly needed scaling up to a more practical model size; at 1/1,200 a destroyer would be about 100mm (4in) in length. This could have led to lots of tricky measurements and calculations but I found that, being in pre-metric times, 1/8in on the plans was more or less equal to 1in at the model's size that I wanted. With care, I could get the basic model sizes and proportions within

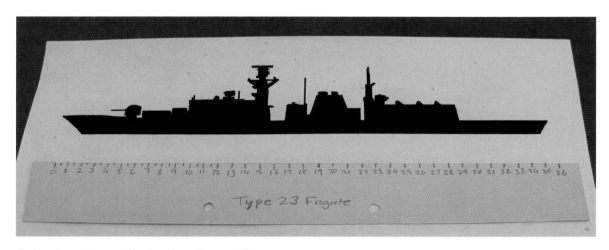

Scale rule against profile drawing of a warship.

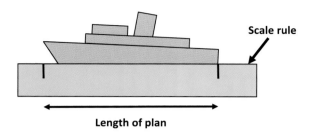

Marking the model length on the scale rule.

a ¼in (6mm), which seemed good enough at the time. Attempting to raise my standards, later models were built from more detailed plans at different scales, so this simple method of scaling up could not be used.

This led me to the adoption of a 'scale rule', a rule marked up so that the model's dimensions can be read directly off the plan/photograph you are using without any calculations. Making an accurate rule with consistent markings was a problem but, like most things, there is an easy way to do it. This starts with a straight-edged piece of card, which is laid against the image you are working on. On to it, the position of the bows and stern are marked.

This length on the card has to be divided up to match the units you plan to use when building the model. Let's make life easy and say that you need to divide this length up into nine equal parts (for most models you will need many more). This could be done with the aid of a calculator or computer, but a simple diagram can be used and re-used many times (mine is currently in its fifth decade).

Taking a suitably large piece of card, a central vertical line is drawn down to a fixed point near the bottom edge. Across the top, a line is drawn perpendicular to the centre line and then equally spaced points are marked on it. The number of points should cater for the model lengths you are likely to need, in this case six units either side of the centreline have been used. From these points, straight lines are drawn down to a fixed bottom point on the centreline to produce a radial pattern.

It is also a good idea to number these lines starting with zero at the centreline.

The piece of card being used as the scale rule is then placed on this ray pattern and, keeping the measuring edge perpendicular to the centreline, moved up and down. The aim is to get the bow and stern marks on the rule to match the number of length units on the ray pattern. In this case it is nine units, so with the bow mark on line number five and the stern mark on number four, the rule can be marked up into nine equal spaces. Using the same method, the units on these scale rules

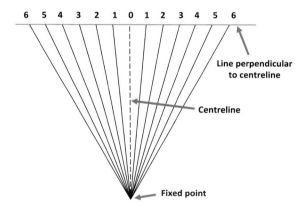

Grid for marking up the scale rule.

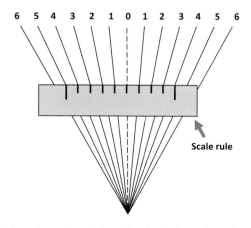

Dividing the scale rule into the desired number of units.

can be further subdivided into halves and quarters, as required.

A scale rule may lack the absolute precision that having drawings or photographs 'blown up' to exactly match the size of your planned model, but with care, it can avoid a model being out of proportion and it is a darn sight cheaper.

LAYOUT SKETCH PLANS

Using sketch plans that show the outlines of the model hull in side and plan views, you can check on the planned structure and positioning of internal structures and items. Plans something like a quarter or half the size of the model can be used, but if trying something for the first time, full-size plans might be safer. Without this stage, it is too easy to find you are unable to install items into the completed model or something that might need replacing or adjusting becomes inaccessible without drastic model surgery.

A sketch shows a typical internal layout in warships. It is, at best, a guide and you can move things around to match your needs. An example was in the Type 23 frigate when it was soon seen that the battery could only be installed through one access opening and this would place it in a compartment ahead of the receiver. This was not a problem and easily sorted before any wood was cut. In other models there would not have been good access for the rudder servo in the stern compartment so, with minor alterations, it could be installed alongside the motor.

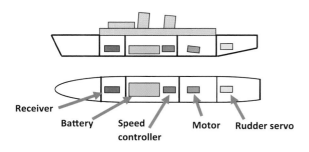

Planning the internal equipment layout.

SELECTING THE BUILDING METHOD

The sketch plan can also suggest the best way to make a model hull from sheets of wood. There are two different ways to do this. Both depend on what parts of the hulls you start with as this determines how the other parts are formed and the hull is built.

Warships usually have a slim hull form to achieve their desired speeds. This can be simulated with a building method based on first cutting out two identical pieces to match the hull sides. Hulls with a fuller hull shape might be better built by starting out with two pieces that match the main hull deck and bottom shapes.

Side Sheet Method

This method of building is shown in a typical cross-section. The hull sides sit on a bottom sheet with the corner junctions reinforced by longitudinal strips. Transverse bulkheads run across the bottom sheet between the two hull sides, but the bulkheads do not reach the top edge of the hull sides. This allows edging strips to run along the top of the bulkheads fixed to the inner surface of the sides.

There is a further recess above these strips. This enables the deck to fit flush inside the top edges of the hull sides. The deck is removable

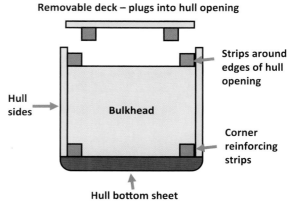

Side sheet method of hull construction.

and is secured into the hull with a frame plug on the underside of the deck. This gives maximum access inside the hull and with an accurately made plug, it is surprisingly resistant to water splashing over the deck. There is also the bonus that when working inside the hull, removing the decks places much of the delicate details safely out of harm's way.

The shape of the hull side sheets is taken from the sketch drawing. With many warships, the edges of these sheets can be simple straight lines with a flat bottom edge until it sweeps upwards at the stern. Note that the depth of the sides must be reduced to allow for the thickness of the hull bottom sheet, and the two side bottom reinforcing strips match the length of the side sheets from the bottom of the bows to where the hull sweeps upwards.

The sides of the amidships section of most warships are parallel or more or less so, but to create the bow and stern shapes, the side sheets will have to bend inwards. They will meet at the bows, possibly also at the stern, but usually the stern is rounded or simply cut off sharply.

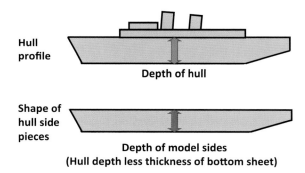

Determining the side shape from the side view.

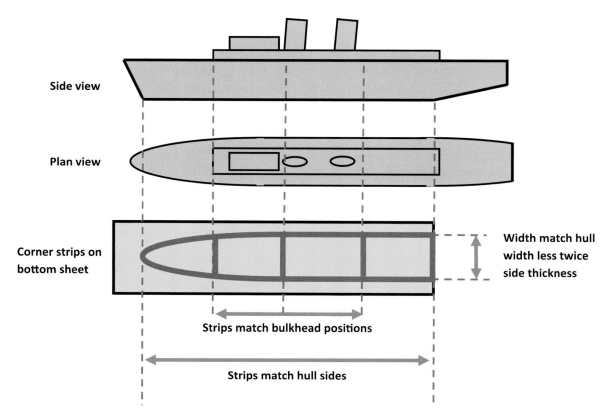

Working out the hull base and bulkhead positions and sizes.

As the bulkheads have to fit inside the hull side sheets, their width is the width of the model at this position, less twice the thickness of the side sheets. Likewise, the bulkhead heights have to match the height of the hull sides at their position less the size of the strip glued to the top of the bulkhead/sides and the thickness of the deck sheet. Notches to accommodate the bottom/side reinforcing strips will also be needed at the two lower corners of each bulkhead.

With the hull parts cut out it makes sense to check that they will fit at the right positions. The longitudinal reinforcing strips match the lower edge of the hull sides and the bulkhead notches match these strips. Very important is that the bulkheads match the hull sides from the bottom edge but stop before reaching the top to allow space for the edging strip and deck sheet to fit.

Plan Sheet Method

In this method of hull construction, the key parts are the hull deck and bottom pieces. These are used to determine the shape of bulkheads and, to make the bow shape, a stempiece. When glued together they make a frame upon which the hull side sheeting is attached. To allow for hull curves, this side sheeting is usually best added in pieces with the wood grain vertical.

Again, a sketch plan of the proposed model is desirable to avoid problems with inconveniently positioned bulkheads plus internal access and placement of items. The deck and bottom pieces do need to be cut to allow for the thickness

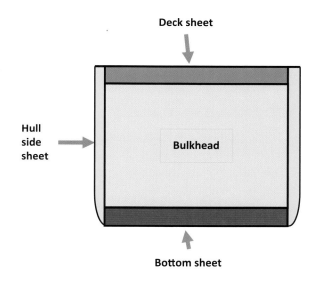

Plan sheet method of hull construction.

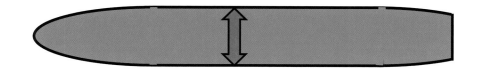

Deck – hull width less twice side thickness

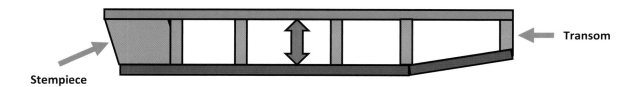

Bulkhead height – hull depth less deck and bottom thickness

Determining sizes and shapes in the plan sheet method.

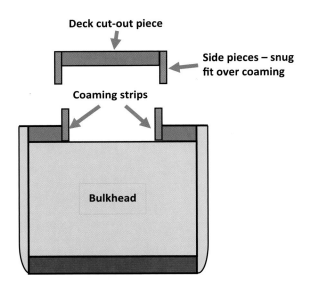

Internal hull access with the plan sheet method.

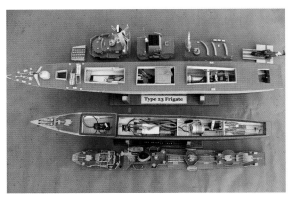

Internal access with the plan (upper) or side (lower) sheet methods.

of the side sheeting and the bulkhead heights to allow for the thickness of the deck and bottom pieces.

Internal access into these hulls can be via sections of the deck being cut away. The cut-out pieces can be used as plugs into the deck openings on the underside of superstructure blocks. A possibly more secure method is to fit coamings around the deck openings and make a close-fitting hatch, which may be a superstructure block or some other deck item

WHICH WAY TO BUILD?

I have suggested that the side sheet method of hull construction favours models based on slim warship hulls, while the plan sheet method is better for 'fuller' hull shapes. In practice, it is possible to use either method to build a model based on many warship types. Even a combination of methods will sometimes prove to be a good answer to any design and construction problems.

Looking at the materials available can dictate the best method to use. The plan sheet method, which uses short lengths of sheet to cover the hull sides, can be an economical way to use up all those pieces too small to use elsewhere and left in the scrap/spares box.

Internal access inside the hull can also be an important factor. If you want easy access to most of the internal spaces then the side sheet method, which allows whole decks to be removed, might be best. If installing items and any maintenance can be done through smaller deck openings, then the plan sheet method is a good option.

Once again, there is no one right way to design these SOS warship models. But, thinking about resources (materials, tools and ability), plus a realistic approach (anything installed can fail), and even if problems arise, you have created the potential to correct them.

Chapter Seven

Building Hulls

Like most tasks that involve making parts and then assembling them, the key to success is doing things in the right order.

To illustrate how to build these SOS hulls by both methods, a simple hull design based on a US Navy destroyer of the World War II period has been used. The 'Gearing' class of destroyers featured a flush deck with only a modest amount of shear that can be omitted. This makes construction simpler without spoiling the model's sailing appearance greatly.

These hulls are about 75cm (30in) in length but these basic construction methods have been used on both larger and smaller models.

Finished destroyer model.

SIDE SHEET METHOD

1. Taking the dimensions from the plan, the hull sides are cut from 3mm (¹/₈in) sheet, remembering to allow for the thickness of the bottom sheet. The bottom sheet is cut from 6mm (¹/₄in) sheet and slightly longer than the bottom edge of the hull sides. A centre-line is drawn on this bottom sheet (a ballpoint pen using light pressure is ideal for this job). The two 6mm (¹/₄in) square strips are cut to match the bottom edge of the hull sides. They should be made from a flexible grade of wood.
2. After marking the bulkhead positions on the hull sides, the bulkheads can be cut out. In this hull it means that bulkheads 2, 3 and 4 have heights to match the sides, less the size of the top edging strip to be fitted around the hull opening and the thickness of the deck sheet; in this example, 6 + 3mm (¹/₄ + ¹/₈in). The first bulkhead is only 3mm (¹/₈in) short

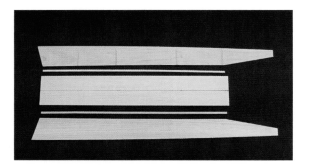

First cuts of sides, bottom and strips.

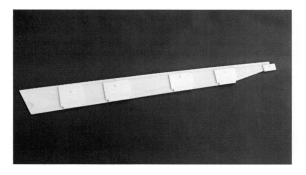

Checking bulkheads against their positions on the sides.

and the last bulkhead is the full height of the hull sides, the bulkhead widths being the desired width of the hull at their locations less twice the thickness of the hull side sheet.

The lower corners of the bulkheads have to be notched to match the strips that reinforce the hull side/bottom junction. In this hull, the last bulkhead (No. 5) did not need these notches.

3. The positions of the bulkheads are marked on the bottom sheet centreline using the hull sides as a guide. In this hull, the fourth bulk-head is at the point where the hull bottom sheet ends. Transverse 6mm (¹/₄in) strips that match the bottom edges of the bulkheads between the notches can then be glued to these positions. The two bottom edge strips are glued to the bottom sheet but only in the section of the hull that has parallel hull sides. In this hull, it was between the second and fourth bulkheads.
4. As this stage involves stressing previously glued joints, it can only be done after the glue has fully set. The bottom edge strips require bending forward of the second bulk-head to meet at the bows, hence the need to use a flexible grade for these strips. To allow the two side sheets to meet at the bows, the ends of these strips have to be trimmed to a triangular section before gluing in place. Pins and a suitable clamp will hold them while the glue sets.

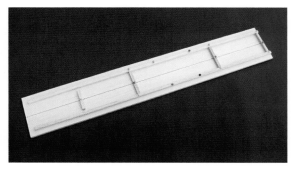

Gluing the strips on to the bottom sheet, parallel section first.

5. It has always seemed a good idea to have a 'dry run' at assembling the bulkheads and hull sides on to the hull bottom using a few pins. If there are any mistakes it is better to find out before any glue is used. Note that both sides, when pulled together forward of the second bulkhead, should meet correctly at the bows.

   If happy, the bulkheads can be glued and pinned in place. A set square is useful to check that the bulkheads are being held vertical while the glue sets.

6. The first hull side is glued to the bulkheads, hull bottom and the strips, but only in the parallel section of the hull. As stated before, in this hull this is between the second and fourth bulkheads.

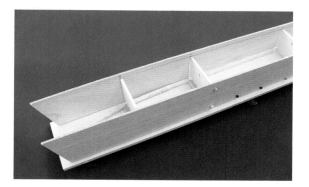

Adding the second side and leaving until the glue has set.

Bending, gluing and pinning the strips in the bow section. Note the joint where they meet.

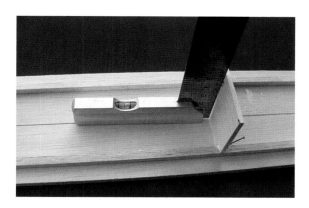

Gluing the bulkheads to the hull bottom.

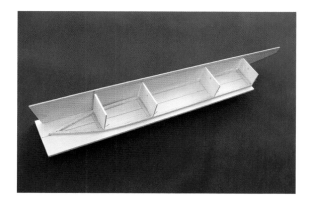

Gluing the first side to the parallel section of the hull bottom and bulkheads.

7. The second hull side is added in the same fashion and the structure has to be left for the glue to fully set.

8. The bow shape is formed by pulling the two hull sides together after applying glue to the bottom/strip, bulkhead and where the sides meet at the bows. A suitable clamp or two at the bows will keep the sides together, but too much force should not be used. This is to avoid pushing the sides inwards and creating a most unrealistic-looking concave bow section. Pins through the hull sides into the hull bottom and reinforcing strips will keep this joint together while the glue sets.

9. As the bows are a potential area of damage when sailing, they should be reinforced. A piece

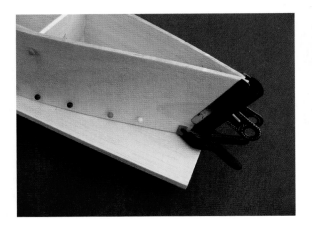

Pulling the sides together at the bows and gluing together.

Bows internally reinforced with glue and tape/gauze.

of glue-soaked tape or gauze, pressed across the inside of the junction between the two hull sides will do this. So far, this technique has survived my numerous accidents!

10. The last bulkhead is glued between the hull sides at the stern. As this involves only slight bending of the sides, a few pins are usually enough to hold things in place while the glue sets. One thing to check is that the sides curve symmetrically or a bent hull will be produced.

11. To match the rest of the hull construction, strips are glued to the inner faces of the hull sides between the fourth and fifth bulkheads.

12. The rear hull bottom piece is glued in place. Because of the angle it makes when it butts up with the main bottom sheet, sanding a suitable chamfer before gluing will make a neater joint.

13. The top edging strips that run around the inside of the hull opening are added next. They fit on top of bulkheads 2, 3 and 4 but leave a recess at the top of the hull sides for the deck to fit inside. Strips also have to be glued between these strips across bulkheads 1 and 5.

14. The hull sides between the bulkheads can be stiffened and strengthened by gluing extra vertical strips between the top and bottom

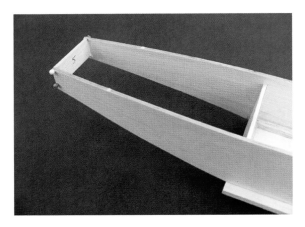

The rearmost bulkhead pinned and glued in place.

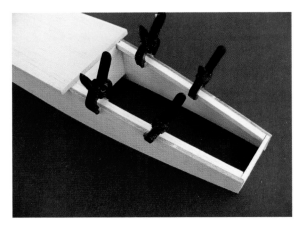

Corner strips glued to hull sides between last two bulkheads.

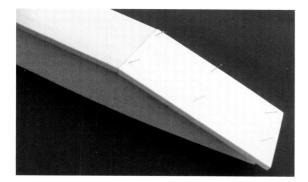

Bottom piece glued to stern on hull.

Razor plane used to shape hull bottom corner.

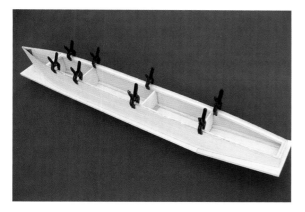

Edging strips are glued inside hull access opening.

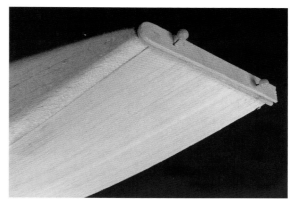

Hardwood strips glued to bows.

edging strips. One such strip between each bulkhead is enough in this size of hull.

15. The hull bottom sheets need trimming back flush with the hull sides. Careful work with a sharp knife can remove the bulk of this material. To avoid the risk of cutting into the hull sides, the knife should leave 2–3mm of wood that can be safely removed with a sanding block.

16. The lower corners of the hull have to be given a curved section (bilge curve). This could be carved with a knife, but a neater and safer way is to use a small razor plane. The edges of the hull bottom sheet are first cut to produce a 45-degree chamfer, after which it is sanded into a curved section. The internal strips along this joint will ensure that no hull strength is lost.

17. Balsa alone would be vulnerable to impact damage in the bows. This is avoided or, to be honest, in my case minimised, by adding a strip of tougher wood to the bows. The joint between the hull sides is sanded just enough to create a flat, gluing area for this strip. It could be a single piece of suitably sized wood, but one laminated from 'lollypop sticks' was used on this hull. Laminating this reinforcement piece from thinner strips is much easier if the bows have a curved shape.

18. After the glue has set and any pins removed, the bow strip can be shaped to blend into the hull. A sharp edge is not needed, which would be prone to easy damage and paint would not adhere well. A small radius will cut through the water easily.

Bow strips carved/sanded to blend into hull shape.

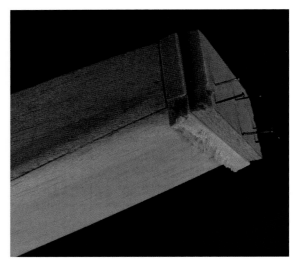

Balsa sheet glued to last bulkhead then sanded into hull shape.

19. The stern of these destroyers was not the flat transom-type but had curved corners. This was created by gluing a couple of pieces of balsa to the fifth bulkhead and then sanding to the desired shape.

20. A frame that fits under the removable section of the deck and plugs into the hull opening can be made at this point. It is constructed from square strips and is built inside the top edging strips. Clamps will hold the longitudinal strips in place while the transverse strips are glued in place.

21. Provided you avoid sticking this frame to the fixed edging strips and bulkheads, it ought to lift out of the hull opening. If it seems tight, then a light sanding of its edges might be called for. One important tip is to mark the top of the frame, as it is unlikely that it will fit into the hull opening the other way.

22. The top surface of the frame is glued to the underside of the sheet, which will become the removable deck section.

23. The deck is offered to the hull opening and will start to fit into place, but excess sheet around the edges will stop it fitting inside the hull sides. This has to be removed by a process of trimming the deck edges a little and refitting. This will almost certainly have to be repeated several times until the deck fits into the hull with a small gap (about 1mm) around its edges.

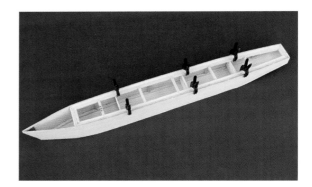

Fitting and gluing together the frame.

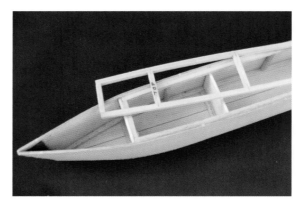

Frame lifted out of hull – note top marked.

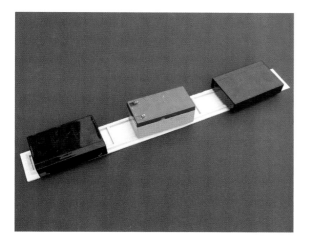

Weights used to hold frame to underside of deck sheet while glue sets.

Fixed section of deck cut and trimmed to fit into place.

Oversize deck fitted into hull opening before trimming.

If, during this process, the deck becomes stuck, then gentle leverage ought to lift it free.

24. The fixed deck from the bows to the first bulkhead can also be made. Again, a 'cut and try' method can quickly produce a shape that will fit within the hull sides. This time, a less than perfect fit can be tolerated since it will be glued into place and any gaps around the edges can be filled.

It is suggested that this part is not in fact glued in place at this stage. Access into the bow section of the hull might be welcome to allow the placement of ballast during early floatation and trimming checks.

PLAN SHEET METHOD

1. This method of hull construction requires a slightly thicker hull bottom sheet and 9mm ($^3/_8$ in) was used. The deck sheet was made from 6mm ($^1/_4$ in) sheet.
2. Starting with the same type of model plan used with the side sheet method, the deck and hull bottom parts need to be drawn on to balsa sheets. These have to be the width of the hull less the thickness of the sheets that are going to cover the hull sides. To produce symmetrical shapes, a simple template with the curved edges of the deck at the bows and stern can be used. Only one side of the hull shape is needed on the template – it can be flipped over when marking out the other side. This ensures that the same curve is made on both sides of the hull.

   Unless you plan to use the template many times, stiff card is a suitable material for this item.

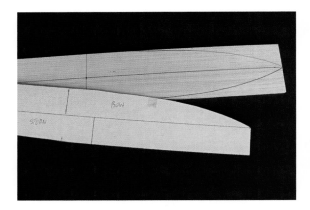

Card template used to ensure symmetrical hull shape.

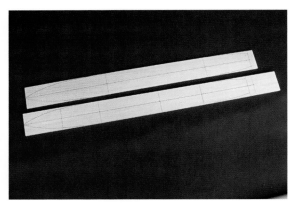

Deck and hull bottom pieces drawn on to balsa sheets.

3. A centreline is drawn down the two sheets and the deck and hull bottom shapes are drawn on. In this model, the deck and bottom bow shapes can be made using one template. But, to create the desired slope in the bows, the hull bottom is shorter than the deck.

4. Using the two hull pieces, the bulkheads can be cut, remembering that their heights match the hull depth less the thicknesses of the deck and bottom sheets. A further complication is the hull sheer (the deck rises on moving from stern to bows), which makes each bulkhead a different height. At least their widths are the same as the deck and bottom sheets.

   A stempiece is need in this method of construction and it runs along the centreline from the bows to the first bulkhead.

   The deck access cut-outs can also be made at this point. This does weaken the deck part a little and there is the option to make the cut-outs after the basic hull construction has been made. The small cut-out at the stern is for access to the rudder tiller. Do keep these cut-out parts as they can be used later.

5. The stempiece and bulkheads are glued to the main hull bottom. Again, a set square is useful for ensuring that they are upright before securing with pins and leaving for the glue to set.

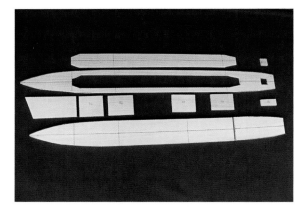

Hull parts cut out before checking the fit correctly.

6. The deck is added to the tops of the stempiece and bulkheads. After a check that the deck sits squarely in place, it is glued and pinned in place.

7. The rear of the hull is completed by adding the last bulkhead and rear hull bottom piece. To make better glued joins, a little chamfering of the top of this bulkhead and where the bottom pieces meet can be used.

8. The edges of this hull framework need sanding to ensure that the hull side sheeting will make a strong joint with it. From the second bulkhead aft probably needs only light sanding. However, due to the slight flare in the bows, more material has to be removed to achieve this.

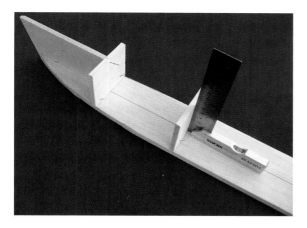

Using a set square to ensure stempiece and bulkheads are vertical.

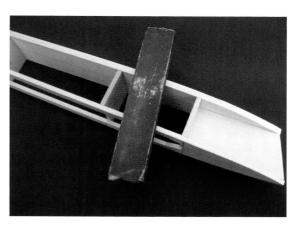

Using a sanding block to shape the edges to give a good surface for the side sheeting.

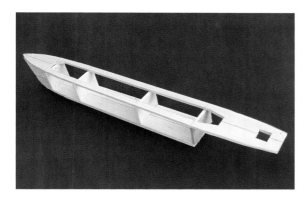

Deck glued to stempiece and bulkheads.

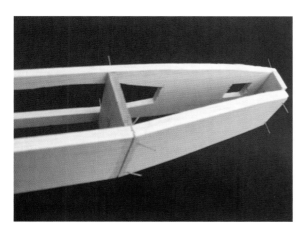

Rear bottom piece and bulkhead glued in place.

A sanding block that covers both the edges of deck and bottom sheets at the same time is the only way to do this.

9. Slightly oversize 3mm (1/8 in) balsa side sheets were added to the hull frame starting in the middle and working on alternate sides towards the bow and stern. These sheets were cut with the grain running vertically between the hull deck and bottom. This makes any curves, in this example in the bow area, much easier to accommodate. Glue is applied to the hull frame and to the butt joint with adjacent pieces then pinned in place.

10. To more readily adapt to the flare in the bow area, a triangular piece of side sheeting can be inserted around the position of the first bulkhead. This was cut so that subsequent sheets would lie with the wood grain lined up with the bows.

11. The bows are best covered with one side sheet at a time. An oversized piece can be glued to one side then pinned to the deck, hull bottom and stempiece. When set, the bulk of the sheet extending beyond the stempiece can be cut away to allow the other side to be completed.

12. The excess side sheeting that extends beyond the deck and hull bottom has to be removed. The bulk can be cut away but care is needed –

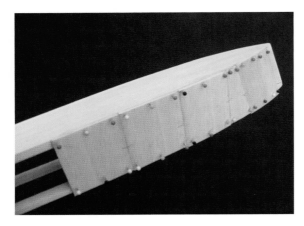

Gluing the hull sides to the hull.

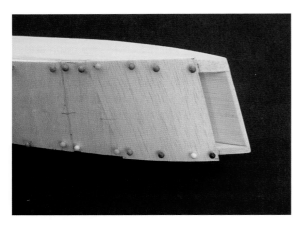

A triangular insert makes it easier for the sheets to follow the flare in the bows.

a sharp blade and cutting from the outer surface inwards will avoid any risk of the balsa tearing.

The deck edge needs a flush finish and the use of sanding blocks, coarse, medium then fine grades, is the safest way to achieve it. The hull side–bottom junction has to have a curved section. A good way to do this is first produce a 45-degree angle at this position with a coarse sanding block, taking care not to remove too much material and thereby weaken this joint. This can then be blended into a smooth, curved section.

13. Hull construction is finished by sanding a flat on the bows, then gluing the hardwood reinforcing strips into place. The curved stern is made from a block of balsa glued to the rear bulkhead. After the glue has fully set, these bow and stern pieces are carved and sanded to blend into the hull shape.

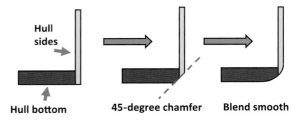

Creating the curve in the hull.

Chapter Eight

Different Hull Forms

Hull construction so far described has been about as simple as you can get. Shapes that are more complex can still be made by using modest changes. An example is where the hull features a raised forecastle, which is the hull being one deck higher in the bow section than aft. In the side sheet method of hull construction, this is produced by cutting the side sheets with the required step. A bulkhead at this position will maintain hull strength. When building using the plan sheet method, the deck has a transverse cut at the step position and the forward bulkheads are taller than the aft bulkheads.

The flare can be produced by changing the third and fourth stages in the description of the side sheet method of building the hull. Rather than have the bottom edge strips curve smoothly to meet at the bows, they are cut (quite often at the step in the hull sides) and glued in a straight line to meet at a point at the bows.

Simple, rectangular bulkheads will not work in this bow area and they need to be wider at the top to match the required deck width and narrower at the bottom to fit the bottom edge strips. The vertical edges of these bulkheads could be straight, but a smooth concave edge can produce a better final hull shape.

The hull sides are added first in the parallel hull section and must be fully set before the next stressful stage. This involves bending the hull sides in to meet at the bows. The bottom edge of the sides requires some persuasion to fit against the straight bottom edging strips, since the top edge still has a curve. I have found that wood strips pushed against the hull sides at this position with a suitable clamp will achieve this. Using medium and hence flexible, but not weak, grades of balsa, will allow the hull sides to cooperate in creating the desired shape. If they are reluctant to make a tight bend where the edging strips change direction, then a *light* vertical scoring, no more than 1mm deep and 10mm high, will allow the balsa to crack but not break apart, and will make the sharp bend needed. Any worries over weakening the hull are avoided with internal reinforcing strips between the bulkhead and sides.

If scoring the wood worries you, then locally wetting the hull sides in this area will soften and slightly expand the balsa and make it more amenable to bending. A damp tissue placed against the hull sides is a better idea than flooding it with water.

In the plan sheet type of hull construction, flare can be created by making the hull bottom

Bulkhead at step in decks

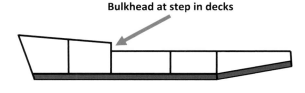

Raised forecastle in side sheet hulls.

Flare created with angled edge strips on hull bottom sheet.

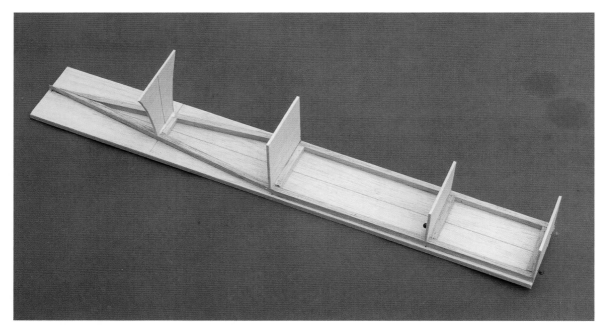

Concave bulkhead edges used to give side sheets flare.

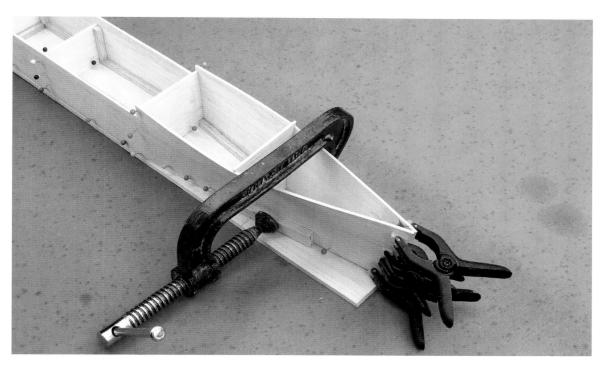

Gentle but firm persuasion is needed to make the sides conform.

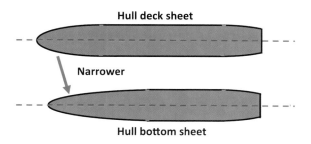

Flare can be made with the bottom sheet narrower than the deck sheet.

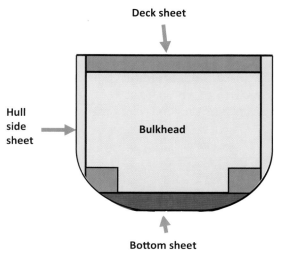

Doublers allow a more generous bilge curve to be made.

narrower than the deck sheet in the bows. Adding the side sheeting in short sections with the wood grain running vertical ought to accommodate the modest curves and twists needed.

One problem with the plan sheet method is that the hull's bilge curve (at the bottom corners of the hull) is limited by the amount of material that can be removed before the joint is weakened. A shapelier hull section can be made by adding doublers to the edges of the bottom sheet.

These can be made by cutting the hull bottom sheet out and then using it as a template to cut the doublers from the remaining balsa sheet.

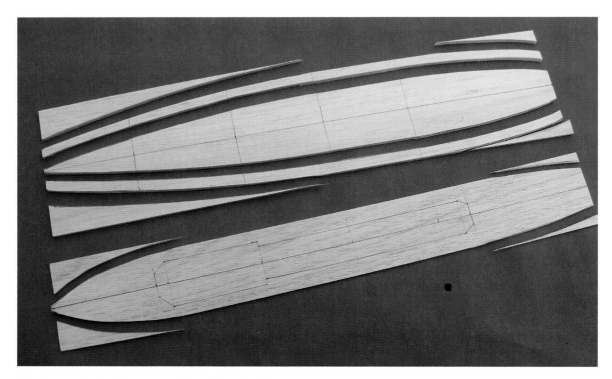

Cutting out the hull bottom and matching doublers.

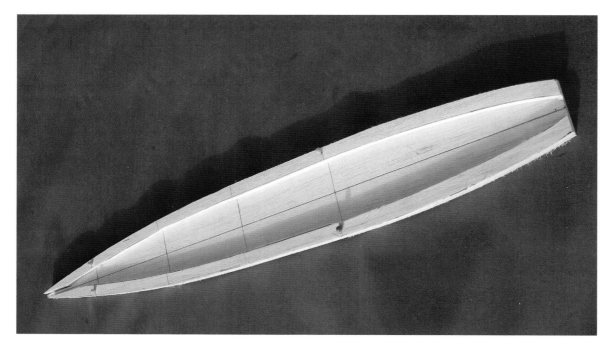

Doublers glued to edge of hull bottom.

The doublers are glued to the edges of the bottom sheet before adding the stempiece, bulkheads and deck.

Pronounced hull shear can be easily added using the plan sheet method by making the height of the bulkheads and stempieces increase appropriately when moving from the stern to the bows. This means that the deck sheet has to bend but not the bottom sheet or a 'banana'-shaped hull will result. This means that a thinner balsa sheet has to be used for the deck, say 5mm (3/16in) compared with 6mm (1/4in) or thicker on the bottom. The edge doublers also help to stiffen the hull bottom. Working on a flat surface and using suitable and carefully positioned weights does the job.

Hull shear is not so simple when using the side sheet method, if the removable deck sections have to be flat. It is possible to cheat a little and break the deck edge of the hull sides in to a few straight sections. This, combined with the hull sides being curved in plan view, creates the right sort of illusion at a distance and allows the flat, removable sections of the deck to be easily made.

Some vessels have even more shapely hulls in the bow and stern areas, and greatly simplified hull sections would not produce the right

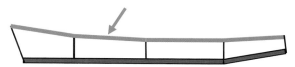

Deck flat between bulkheads

Hull shear suggested with flat deck sections.

appearance. A model based on the US Navy's seaplane tenders used in World War II was one. They were converted from merchant ship hulls and did not feature the simpler high-speed hulls found on most warships. As a result, the plan sheet method was used, with doublers along the edges of the hull bottom sheet. This enabled the lower sections of the hull to have the desired concave shape in the bow and stern. The stern also required an extra bottom sheet running from the penultimate bulkhead to the stern.

After sheeting the sides, sanding the corners, general filling and shaping, the resulting hull was a close match to the original vessels. In fact, quite a few people would not believe that it was built from balsa sheets until they looked inside the hull!

Another challenging construction was the bows of a model based upon the 'Colossus' class of light

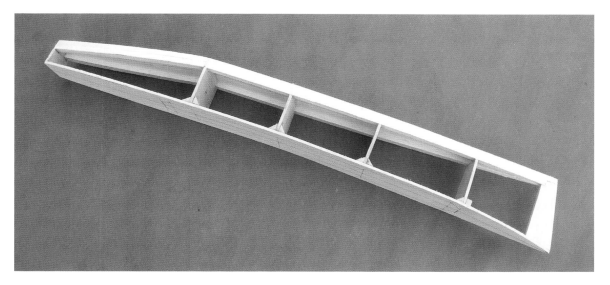

Curved deck added to plan sheet hull.

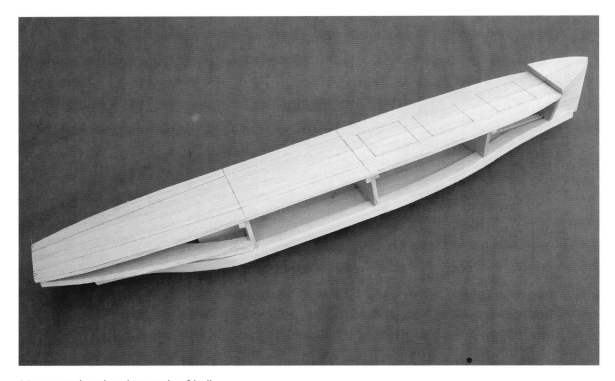

More complex plan sheet style of hull.

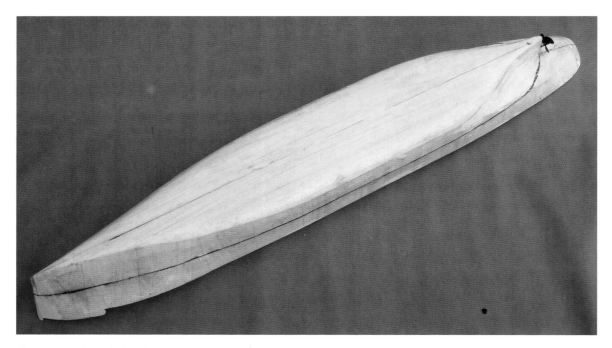

Shapely hull from balsa sheets.

fleet carriers built for the Royal Navy. They have a characteristic 'streamlined' bow shape with pronounced flare that, realistically, could not be omitted.

Using the side sheet method of hull construction, along with lite ply sides, the bow shape was cut out including the radiused leading ledge of the flight deck. The lower part of the hull was built as per the side sheet method with the hull sides meeting at the bows. The bow shape was started by gluing a highly curved former between the hull sides.

The flared section of the hull sides was formed from pieces of thin ply sheet. This would accommodate the flexing to produce the bends and curves required for gluing to the hull. The final bow shape was made from balsa laminations.

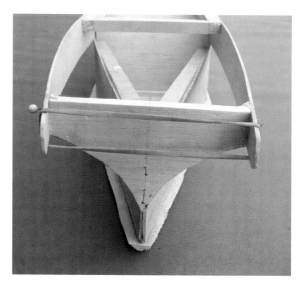

Starting to create the bow shape.

Highly curved bow sections in this aircraft carrier model.

Carrier bows ready for final sanding.

TYPE 23 HYBRID

The model based upon the Type 23 frigates ended up being something of a combination of different methods of hull construction. It had to incorporate the sinuous deck curve along with obvious flare in the hull sides. The hull was started in the side sheet style with bottom edging strips glued to the hull bottom sheet with parallel sides amidships but running straight to the bow junction. To allow for a generous bilge curve, the bottom sheet was 9mm ($^3/_8$in) thick, while the strips were 12mm ($^1/_2$in) square.

Construction then veered towards the plan sheet method. The bulkheads were cut with their vertical sides angled outwards to create the hull flare. The stempiece had to duplicate the Type 23's

characteristic highly curved bows. After gluing them on to the hull bottom, the bulk of the excess bottom sheet was cut away.

The deck had to be cut from 4mm ($^3/_{16}$in) balsa sheet to accommodate the sheer in the hull, which reverses as it approaches the bows. Gluing the deck required careful positioning, some encouragement from suitable weights and a few pins to prevent any movement.

The hull structure was completed with the addition of the transom and the rear hull bottom piece. Then, in the plan sheet style, the edges of the bottom and deck were sanded to allow sound glued joints with the side sheeting.

The balsa sheeting on the hull sides had to accommodate some bending and twisting to match the hull shape. Rather than the usual plan

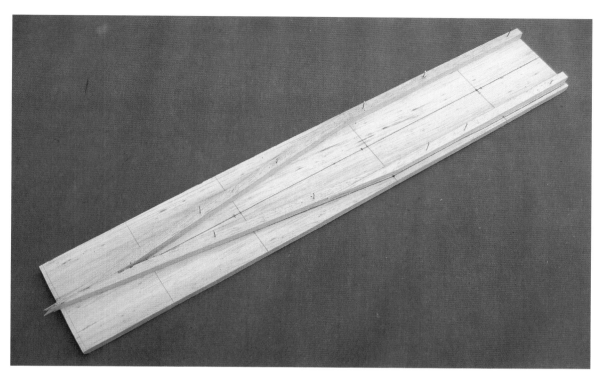

Type 23 frigate hull bottom with substantial strips to allow for a generous bilge curve to be carved.

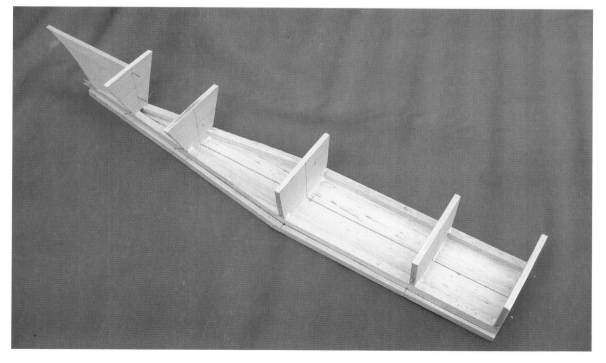

Stempiece and bulkheads glued to hull bottom.

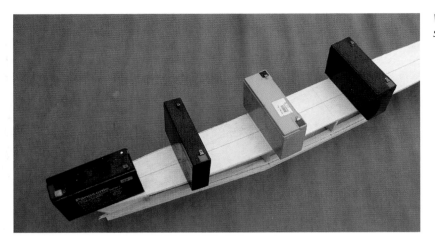

Weights used to maintain deck shape while glue sets.

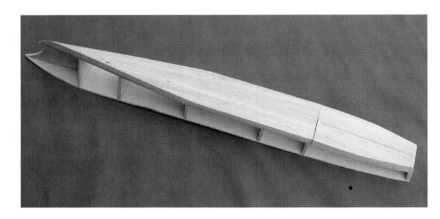

Edges prepared for the addition of side sheets.

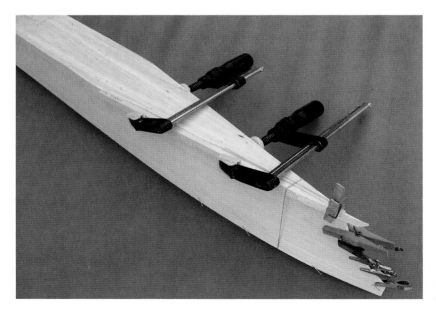

Side sheets added with some bending and twisting.

sheet method of using balsa with the grain vertical, covering the hull sides with the grain running lengthways worked better. Two sheets were glued to the hull, first to the parallel amidships section, before being pulled inwards to create the bow and stern sections. In the bows, the sheets required some persuasion with clamps, but pins and elastic bands were strong enough in the stern.

After trimming the excess sheet away from the deck and bottom edges, the bow reinforcement was added. Using laminations of 'lollypop sticks' coped with the curved bow shape. The bilge curve between the bottom and side sheets was started with use of a razor plane followed by sanding blocks

Due to the more complex deck sheer, this was one model where it seemed better to make the access cut-outs after building the hull. This had been planned for when drawing out the drafts for the model, so all the bulkheads were correctly positioned to avoid any problems.

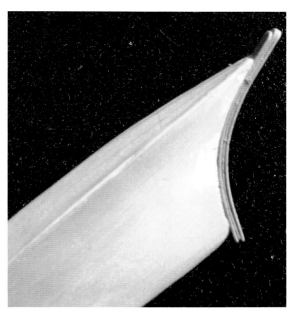

Curved bows reinforced with hardwood strip laminations.

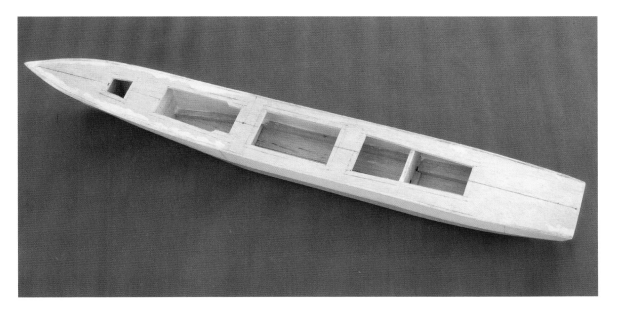

Deck access openings cut after completing hull construction.

Chapter Nine

Internal Outfitting and Test Float

With the model's hull structurally completed, it is a good time to install some of the internal items. This can avoid the risk of damaging other items, such as the superstructure and details, which will be added later.

DRIVELINE

Getting the power from the motor to the propeller requires two more items: the shaft upon which one end has the propeller fitted and a coupling to connect the other end with the motor shaft. As the shaft has to rotate without letting water into the hull, it runs inside a fixed tube passing through the hull.

The shaft usually has a threaded bottom end on to which the propeller is screwed. The type of threads ought to be checked for compatibility – most are now 'M4' type. Just screwing the propeller on would not be secure as it is not unknown for them to unscrew themselves and drop off! For this reason, a lock or jam nut is screwed on to the end of the threaded section first. Then the propeller screwed on and tightened against the lock nut until it is firmly jammed in place. Holding the propeller in your fingers but using a small spanner on the nut should create enough force to ensure that the propeller never departs.

The tube should have a larger internal diameter than that of the shaft but be supported in short, close-fitting bearings at each end. This is to minimise friction that a full-length, close-fitting tube would create. A small washer, sized to fit on the shaft, is often inserted between the lock nut and the lower tube bearing. The washer can be metal

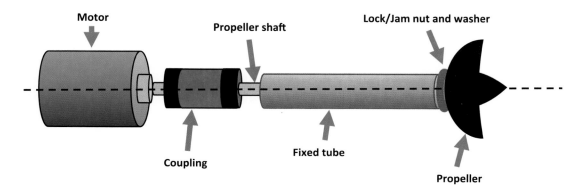

Driveline components.

or a hardwearing plastic. It acts to further reduce friction and to make a more watertight seal when the propeller rotates and presses the washer on to the lower bearing.

Perfect operation of the driveline requires the motor and propeller shafts to be exactly in line. This is not easy to achieve and this is where the coupling comes in. There are different types, some based upon the Universal or Cardan/Hooke coupling, others on a ball-and-socket type. These will allow for angular misalignment but only if the two shafts are in the same plane. Doubling up these couplings, essentially two couplings 'back to back', can accommodate more misalignment but, to be honest, no coupling should be used as a substitute for getting both motor and propeller shafts as closely aligned as possible.

A surprisingly effective coupling on lower powered models can be a flexible tube, which is a tight fit, on to the motor and propeller shafts. Rubber tubing can perish and fail, and some plastics are too stiff, wasting power unless perfectly aligned. Silicone rubber tubing is better as it is flexible, strong and long-lasting. In low-powered installations, an alternative could be the thick-walled but still flexible insulation stripped from electrical wires.

A problem with tubing can be creating a secure grip on the shafts, especially when they have different diameters. One option in this situation is to use a smaller tube that will grip the shaft as an insert in the larger tube that transmits the power. Alternatively, the metal inserts used in the other types of couplings (which will fit on to

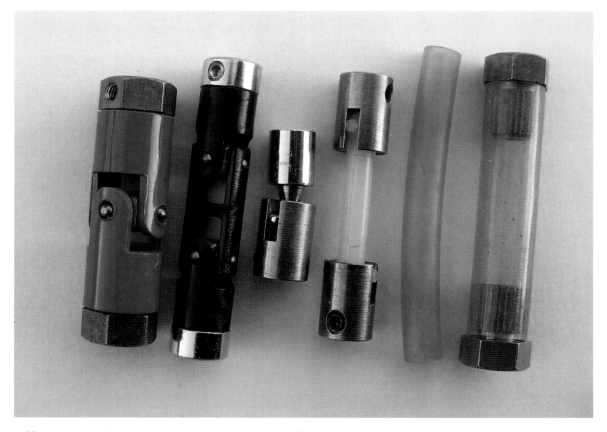

Different types of couplings between motor and propeller shafts.

the shafts securely) can be connected with tubing that would otherwise be much too large for the shafts.

SINGLE OR MORE?

It is fair to say that most full-size warships will employ more than one propeller. It might seem that any model based upon them ought to have the same number of propellers too. This is certainly true if the model is being built to spend its life a full hull display model but not necessarily so if it is to be a practical working model boat. A single, well-matched motor, propeller and rudder will usually be the most economical and reliable way to achieve good performance and handling. In addition, for a first foray into designing and building your own model, the complication of multiple motors and propellers is asking for extra problems.

MOTOR MOUNTING

Motors can be attached to mounts (either plastic or metal), which are then secured into the hull. This is usually done by securing the mount with screws into a wooden base; obviously, by using screws that are big enough to do the job but *not* so long that they puncture through the hull bottom. As the propeller shaft is likely to be angled downwards, a wedge-shaped block may be needed under the mount to keep the motor and propeller shafts in line.

If the motor is a cylindrical body with no separate mount, then other methods can be used. It is possible to hold the motor between two wood blocks, with a securing strap across the top and screwed to the blocks. To prevent the motor moving, placing some thin rubber strips between the motor body and the securing strap is very effective.

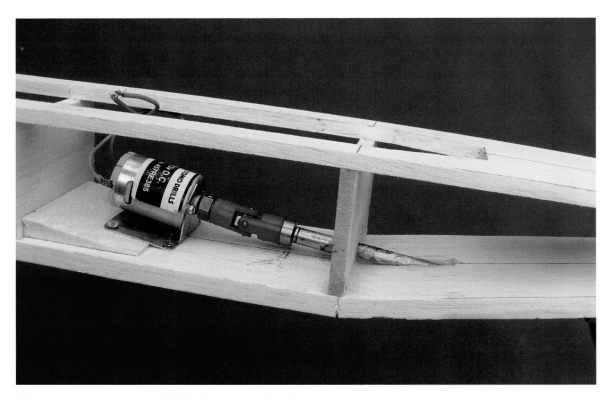

Motor mount secured to angled block to get shafts aligned.

Motor secured by strap between blocks.

To be honest, with the modest power of motors like the RE360/385 types it can be sufficient to stick the motor into the hull, using one of those 'all-purpose' adhesives that claim to bond just about everything will work, although a silicone sealant might be safer if you ever want to remove the motor without damage to the model. It is also essential to ensure that the motor case is clean and grease-free, otherwise you risk it coming loose, probably at the worst moment possible! Whatever is used, the motor and propeller shafts still need to be aligned and some packing under the motor might be needed.

INSTALLING THE DRIVELINE INTO THE HULL

Drawing a centreline down the hull bottom, where the propeller and rudder tubes are to fit, is the first thing to do. Placing the rudder tube first

is a good idea as it avoids the risk of fouling the propeller. There also needs to be a check that the tiller arm will fit below the deck and can rotate without catching anything. It ought to be capable of turning 45 degrees either way.

You can make your own rudder assembly, but unless something out of the ordinary is needed, then a commercial item can be more economical. The two hulls described in Chapter 7 used such rudders, which were comprised of brass blades fixed to brass shafts that slid into plastic tubes. The tubes were threaded so that brass nuts could be used to secure them into the hull. Double-sided plastic tiller arms were clamped to the top of the shaft with small bolts. The rudder blades were of the balanced type with the leading edge being forward of the shaft axis. This reduces the forces needed to turn the rudder, and inter-cepts more of the water being driven backwards from the propeller. The rudder should be deep enough to cover the water flowing off the

Low-power motors can be stuck in place.

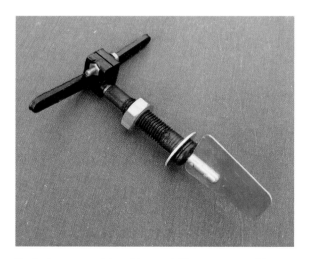

Typical commercial rudder and tiller arm assembly.

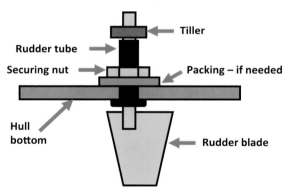

Method of mounting rudder assembly into a hull.

propeller – much larger than this can lead to uncomfortable handling and excessive rolling of the model when the rudder is applied.

When tried in the hulls, both rudder assemblies needed the shaft and tubes reducing to clear the decks, and the tiller arms shortening to allow free movement of 45 degrees each way. The tube was then glued into the hull bottom using epoxy. Brass washers were supplied to fit between the securing nuts and the hull bottoms. As the tubes are not threaded along their whole length, this may leave a small gap between the washer and hull bottom. If this occurs, then some extra packing is needed, either another washer or a piece of wood with a hole made to match the tube. An application of epoxy to the nuts and washers will then make a secure and watertight installation.

The rudder will probably get in the way when installing the propeller shaft tube. It can always be slipped into place to check clearances before sticking anything in place. The location of where this tube passes through the hull bottom sheet is needed. The layout drawing, even if at reduced size, can give you a good idea of where this is. An alternative is to hold the tube against the side of the hull and 'eyeball' things, but remember to allow for the motor, coupling and propeller. Having marked the tube's position on the centreline of the hull bottom sheet, a hole is made vertically through the bottom sheet, which can then be opened out into a slot to accommodate the tube at the correct angle. I have found a 'round bastard file' (one with coarse teeth and close to the tube's

diameter) good for this task, although it needs firm but careful use.

A hole will also be needed in the bulkhead that the tube has to fit through; again, careful work with the file can do this. This is where the plan sheet method has an advantage, as the rudder and driveline can be easily adjusted for alignment before adding the hull side sheets.

It is highly likely that the holes in the hull bottom and bulkheads will need enlarging before everything is in line. It is also essential that the propeller will not contact the rudder and will clear the hull bottom when rotating. A bottom clearance of 3–6mm ($\frac{1}{8}$–$\frac{1}{4}$ in) is adequate – too steep a propeller shaft angle could produce odd sailing characteristics. The position of the tube can be adjusted using wedges or strips of wood where it passes through the bulkhead and hull bottom.

Relative positions of propeller and rudder.

The motor must be at the correct angle to line up with propeller shaft. This can be done with the coupling fitted to both shafts, or with a similar length of tube that is a close fit on both shafts. Both hulls used a piece of balsa carved into a wedge shape to make a base for the motor mounts. There has to be a degree of 'cut/sand and try' before this is achieved.

One tip: when the coupling is in place, apply only limited power to the motor, possibly one or two dry cells, so that it barely runs. To avoid damage, a small spot of oil ought to be applied to the motor and tube bearings; a light oil (like the common '3-in-1' type) is suitable. The best motor position is, when it runs at maximum speed, something easy to see and hear. When happy, the tube and motor base can be glued into place. If the tube requires packing where it passes through the bulkhead, this can be left and the epoxy applied to seal any gaps, as well as securing the tube. Where the tube exits the hull bottom, any packing pieces that stand proud of the hull surface ought to be carefully trimmed flush before epoxy is applied. If any gaps

or depressions are left, they can be filled later with epoxy or a suitable filler. The motor can be screwed into place but a low power recheck for alignment is not a bad idea before going any further.

RUDDER SERVO

The last outfitting job is to install the rudder servo. These are normally very reliable items, but there is always the chance that they could fail and so they need to be accessible, just in case! With the side sheet method, the removable deck gave good access into the hull. The simplest solution to mounting the servo was to screw it to two transverse balsa strips. One was glued across the rear face of the bulkhead and the other between the two hull sides. Before doing this, the servo operation was checked to ensure the linkage to the rudder tiller would not foul anything with the deck in place.

Other features to note here are the use of a 'double servo linkage', that is, both sides of the

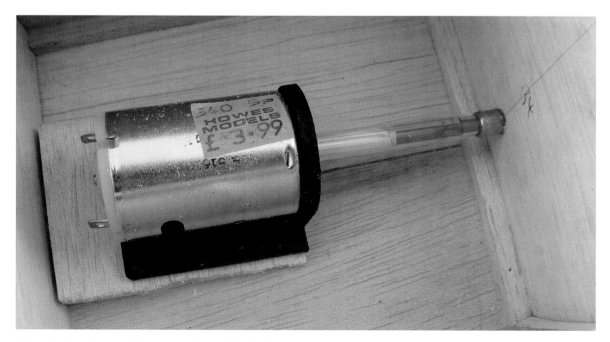

Close-fitting tube to aid motor and propeller shaft alignment.

Double linkage used between servo and tiller arms.

servo arms are connected to both sides of the tiller arms. This creates a very secure push–pull action, provided that the arms are square to the linkage wires. There is no need to have very stiff wire in this system and soft steel was used with a diameter to match the holes in the arms. The connection with the wire link to the tiller arm was made by using the simple and very secure 'Z bend'. You can get special pliers to make these bends in one go or just use suitable pliers to make the bends in two goes. Fitting the linkage wires to the servo arms was via connectors that were fitted to the arms. The wire slides through a hole in the connector and when the correct position is found, a screw firmly secures it to the connector. This is much easier than trying to cut and bend the wire to the exact length.

Having to work through the deck cut-out in the plan sheet hull was not so easy. The servo could have been placed in the same position but access was very restricted. A better solution was to fit the servo in the motor compartment again using

screws into two balsa strips. It looked tight but the servo could be installed and removed with little difficulty. Only a single wire link could be run to the tiller arm, so a hole had to be made in the bulkhead. This wire also had to be stiff enough not to bend as it had to both pull and push when moving the rudder.

Fitting the wire link between the rudder servo and tiller arms was not as easy as it was in the hull constructed using the side sheet method. The tiller was only accessible via a small hole cut in the deck. The best method was to first fit the wire link to the tiller, then thread it through the hole and secure the tiller arm to the rudder shaft.

SURFACE SEALING

It is always good practice to get the hull into the water before starting on all the demanding work needed to complete the model. Knowing that

Less access, so servo placed in motor compartment.

Tighter fit, but tiller arm still accessible.

the model will float on the desired waterline in a stable fashion and that the RC functions correctly is a handy thing to know. It is also not unknown for enthusiasm to flag a little part-way through a model's construction. The sight of even a bare hull moving in its proper element ought to restore the desire to complete it. The other reason is to avoid the situation where the finished model has serious problems when placed in the water. This could result in having to undo and replace a lot of work and, in some cases, this might not be possible. It would still be a learning experience, but hard to call it a good one.

Details of how to seal and waterproof the surfaces of a wooden hull were given in Chapter 3.

Some people advocate sealing both the inside and outer surfaces of a wooden hull. This appears to be sensible, but should water ever penetrate into the wood through cracks, damage or untreated areas, then it will be hard to remove. Damp wood trapped under impervious surface layers can decay unnoticed until serious damage has occurred. For this reason, I only waterproof the outer surfaces of a model. Provided the access points into the hull are reasonably water-resistant, then little if any water should find its way inside. Different techniques are available, but with these two hulls I went down the route of using clear acrylic varnish. Four thin coats, with light sanding between each, produced a tough and watertight surface on the hull.

Careful packing allows waterline to be drawn around hull.

A waterline around the hull is a good idea for checking the model's trim during these first-on-the-water trials. One of the simplest ways to do this is to mark where you want the line to be on the hull and then place the hull on a flat surface. Some packing under the hull might be needed to keep the propeller and rudder clear of this surface. I found a suitable slab of foam plastic for this job. The line is drawn around the hull with a permanent marker pen, which is held horizontally and at the right height with some suitable packing. Sometimes it seems best to keep the hull stationary and move the pen around the hull, and other times to keep the pen still and move the hull – your choice.

INTERNAL LAYOUT

As with most things, there is no universal plan for the internal equipment that will suit all models. The sketch plans drawn up when starting a project ought to have given you some idea of where to place things, but sometimes when you come to do this, changes have to be made. The layout used in this hull was designed to give good access and allow easy adjustments and even replacement should anything fail. Another thing to consider is neat installation of the wires, which ought to be logical (to avoid confused fumbling, especially at the pond side) with no risk of wires fouling anything.

It may be that the leads between the components are too short for your planed layout. Servo extension leads, which have plugs and sockets that match the radio gear, can be fitted between the receivers, servos and ESCs. One aid to keeping things neat is to run leads and wires through holes in the bulkheads. The holes need to be just large enough for the job and should be in the upper half of the bulkhead. Bulkheads are a vital part of the hull during its construction but they also have a safety function. Should the model ever take on water, then bulkheads can limit the internal flooding and give enough time for recovery.

Punching big holes low down will allow adjacent hull compartments to readily flood, which is rarely good news.

INTERNAL BALLAST

Initial calculations suggested that these models would weigh around 1.3kg (3lb) and, at this stage, they were barely half that value when checked on some kitchen scales. Ballast was clearly needed, and to ensure stability it had to be something dense and flat to sit on the hull bottom sheet. Lead or steel are generally the best things to use. Lead can be obtained in thin sheets ('lead flashing' as used in the building trade) and is quite easy to cut to shape with a wood saw or even scissors. If a lead block is the wrong shape, you can even reshape it with hammer. Steel is not so easy to cut, but a vice and hacksaw will do the job.

The ballast pieces were distributed in the hull compartments in an attempt to keep the model's centre of gravity more or less at the amidships position. The ballast was not stuck in place as some on the water adjustments were likely to be needed during the test float and when the model is completed with the superstructures and details added. The battery pack was, however, wedged in place with some foam plastic pieces to prevent it moving during the test float.

TEST FLOAT

The traditional place to give a model its first taste of water is the domestic bath. This is not always possible or acceptable but you do need calm and still water, so the local public sailing waters might not be the best place. Many years ago, I dug out a garden pond, ostensibly for the wildlife that it certainly does attract. However, my family quickly realised it had an alternative use for testing models.

The model was lowered into the water and floated at almost the correct waterline. Some replacement and movement of the ballast pieces had it sitting level and pushing down on one side of the hull (heeling) resulted in it smartly returning upright. If the model fails to show this stability, then now is the time to discover the problem, which will only get worse with the addition of more top weight. In this situation, the model's centre of gravity is too high and it needs lowering. Is it possible to lower the internal ballast or add extra ballast? The only other weighty item is the battery and a change to a smaller one might be possible, which would allow more ballast to be added. If none of these ideas work, then external ballast underneath the hull could be tried. This is a rather inelegant solution but much better than having a delicate model, which makes every sailing session a nerve-racking experience.

Having all the RC gear installed and switched on allows it to be checked for reliable operation. The motor should run smoothly in both directions, but do this only if the bearings (motor and propeller tube) have had a spot of oil. The rudder's movement ought to follow the transmitter controls perfectly with no hesitation. I have to confess that I can never resist some slow manoeuvring around the pond, which does give some feel for how the completed model is likely to handle.

The trial float is also an opportunity to check the motor's current draw. This does involve placing an ammeter in one of the motor leads, securely restraining the model and applying full power via the transmitter control. In this particular model setup, the current was found to be just under 1A and, since the battery was a six-cell pack (7.2V), the power would be around 6W. Knowing the model's weight, as described in Chapter 2, this would put the model's sailing characteristics into the handy, nudging into the lively, category. This seemed appropriate for a model based on a destroyer.

TYPE 23 PROBLEMS

The hull of this model had its internal access via five holes in the deck. In the reduced scale draft

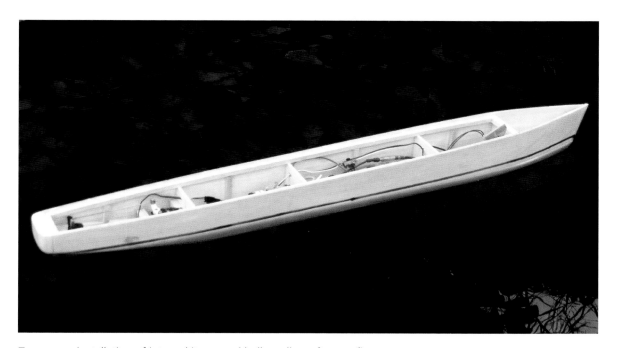

Temporary installation of internal items and ballast allows for test float.

Installation problems solved with a geared motor unit.

plan it appeared that a direct drive from a single motor would easily fit into this layout. However, with the hull construction finished, it became clear that this would not produce an accessible motor installation. Other installations were contemplated but nothing felt comfortable, until a small commercial geared drive unit was tried. This fitted on to the inner end of the propeller tube with the motor mounted above the tube to produce a compact installation. The screws securing the two gearwheels and the motor were easy to reach and it proved successful.

The internal layout had to be different with the battery pack installed ahead of the receiver and ESC. The rudder servo could be fitted into the rear compartment with a double linkage to the tiller arms. The tiller was more readily accessible, as a large deck cut-out was possible under the helicopter landing pad. This goes to show that the internal placement of items is not usually critical, but a tidy and reliable one is essential.

Larger tiller access opening under helicopter landing pad.

Chapter Ten

Sides and Superstructures

WOOD LOOKING LIKE STEEL

The sides of a metal warship are never perfectly smooth. If built by riveting steel plates together, the overlap between adjacent plates can be an obvious feature. Even all-welded hulls usually show the run of the plates. Careful painting can simulate these effects, but this requires exceptional skill and patience. An easier technique is to 'plate' the hull sides with thin card of a thickness you find used in greeting cards.

If you have details of the actual vessel's construction, then you can find the sizes, shapes and location of the real plates. If not then a good idea can be found from similar vessels and will probably not be far wrong. A confession here is that I only plate my model hulls down to the waterline. The justification is that you cannot see the hull below the waterline when it is sailing and it avoids some very tricky curves.

When sticking individual plates on to the hull with a slight overlap, starting from the stern has always proved to be the best method. The overlap faces rearwards and the water flowing past the hull cannot lift any edges. This does not work when moving astern, of course, but most people do not spend much of their time sailing in this direction. The best glue for sticking card to a hull is an impact adhesive and rather than letting the

glue go tacky to produce an instant bond, it is safer to immediately press the card in place to give you some 'shuffling time'.

With the smoother, welded type of hull, you could just cover the hull sides in one piece. However, sometimes a distinctive weld line or run of the plates can be seen along the hull sides. This can be suggested by gluing strips of a suitable width along the length of the hull but leaving a gap between them. If the hull has noticeable sheer then these strips should follow the line of the deck edge. Following my aversion to adding below the waterline, this often means the bottom strips have to be trimmed to match the waterline.

The card can then be sealed in the same fashion as the hull. Two or three light coats are usually sufficient, with very gentle sanding to remove any roughness at the card edges. The first coats of sealant will penetrate into the card, cover any small gaps at the edges and bond it surprisingly well on to the hull. There is a secondary benefit of doing this as the card, despite its thinness, will strengthen the hull and add a useful extra scratch and ding resistance.

An alternative to paper-based card plates would be to use thin plastic sheet. This would avoid the need to waterproof the plates but care is needed, especially on curved surfaces, to ensure that they are firmly bonded to the hull.

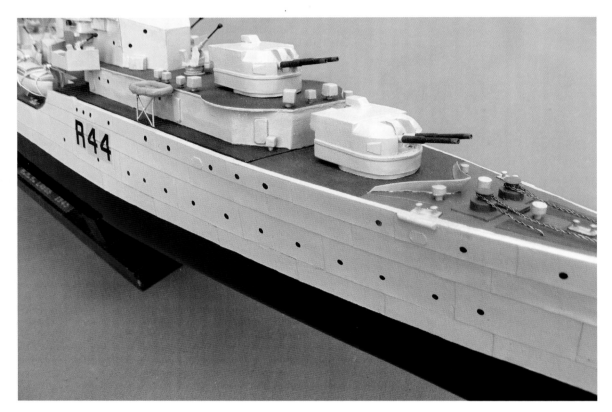

Card plates stuck to the above-waterline hull sides.

Card strips glued along hull sides to suggest welded hull.

APPEARANCE MATTERS

The hulls of warships, having to operate in the same basic fashion, will tend to follow similar design parameters. Not so with the superstructures, which can vary drastically between nationalities, time periods and operational demands. If a model is to effectively represent a vessel that really existed, then all the above-deck structures must create the right impression. Some degree of simplification can be tolerated, possibly hidden from view with a little cunning, but the basic shapes and proportions must still look right.

Mixed proportions can be a trap for the unwary to fall into. You might see a model that looks like it was built by more than one person, possibly all working to high standards, but noticeably different scales. No matter how much effort has gone into such models, they never look right. One tip is to have a 'figure' available that matches the scale of the model. This can be an actual figure, for example those sold for 'N gauge' model railway layouts are 1/160 scale and work well at the nearby 1/144 scale. Alternatively, it could just be a piece of material that matches the human shape at this scale. These 'figures' can be moved around the model to check that spaces, fittings and openings are suitable for the crew.

UNWANTED WEIGHT

It cannot be repeated too many times that everything above deck level in a warship model must be as light as possible. Any weight added here would be many times higher above the waterline than any ballast inside the hull is below the waterline. A top-heavy model can be hard, if not impossible, to return to a stable state with just internal ballast. It is not unknown to see modellers having to resort to external ballast beneath the hull before daring to risk sailing their creations.

Luckily, on models, the superstructures are usually just for appearance with no great structural

or functional demands placed upon them. This means light and simple building methods will work and, even if the hull was not built from wood, it is not a bad idea to make the superstructure using wood. Balsa is ideal for this job as its low density means that robust thick sections can be used without any weight penalty.

LAYOUT PLAN

If the superstructures are to be built on to a flat, removable deck, as in the side sheet method of construction, then the decks should be sealed before adding the superstructure. This avoids having to work around awkward corners and in tight spaces. A centreline down the decks can be useful when drawing the outlines of the superstructure blocks. Do note that these blocks are not always symmetrical.

These blocks can be built up from balsa strips, cut and glued together to create the superstructure outline. This calls for the strips to match the desired superstructure height, less the thickness of the material used to make the superstructure decks. Balsa, lite ply or even card (but a little thicker than that used to plate the hull sides) can be used for these decks.

With the low density of balsa, making parts of the superstructure from solid wood is quite practical. To get the right thickness, the blocks can be laminated from thinner pieces, which can be a good way to use up otherwise scrap balsa. With flat decks, superstructures can be built directly on to the deck, but with overhangs for things like gun positions and walkways, it can be better to build them off the model, add the deck and then glue to the main deck.

Whichever method is used, to avoid any unrealistic sight of wood grain in what should represent a metal superstructure, the vertical surfaces of wood can be covered with thin card. It might also hide any less than perfect glued joints.

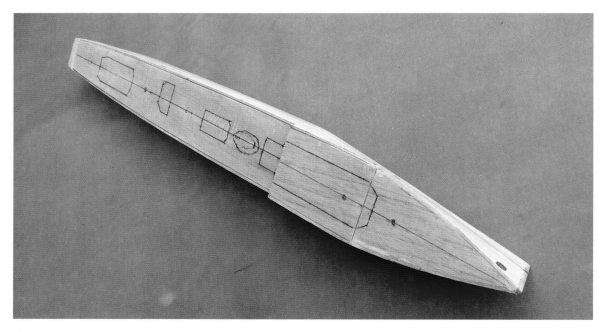

Centreline and superstructure positions drawn on to deck.

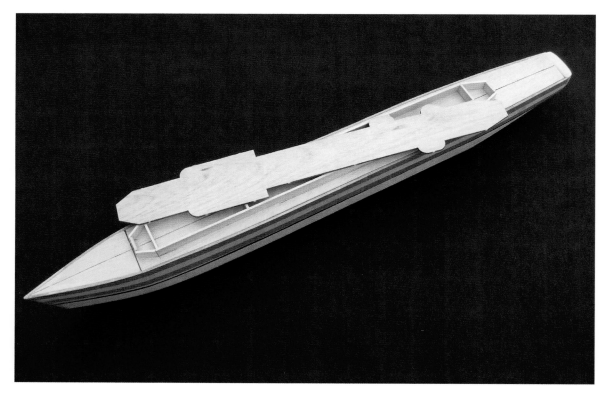

Superstructure outline made from balsa strips and deck about to be added.

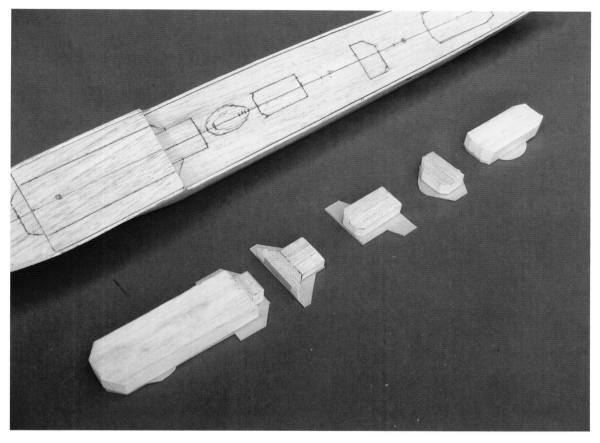

Balsa block superstructures, decks added before sticking to the hull.

DECK COAMINGS

If the hull has been built in the plan sheet method, then access into the hull can be by detachable superstructure blocks or hatches. Making these items fit over vertical coaming strips glued around the inner faces of the deck openings will make an easy opening, yet secure and water-resistant fitting. One tip is to use the piece cut out of the deck to make the top of the detachable piece.

The coaming strips need to extend something like 10–15mm (about ½ in) above the deck level to create a snug fit with the superstructure. On smaller models, card can make adequate coaming strips and around 2–3 mm ($^3/_{32}$ – $^1/_8$ in) thickness is

suggested; alternatively, balsa strips can be used. But, whatever is used, it must be installed with no gaps between coaming and deck, and then thoroughly waterproofed.

To ensure a good fit with the deck coaming, the superstructure is best built over the coamings. This does run the risk of you accidentally gluing it to the coamings and deck, but being aware of this problem usually makes you careful enough to avoid doing it. If it is a serious risk, then a layer of thin plastic (the clear film used in many kitchens is ideal) can be used as a protective barrier. However, care is now needed to avoid this plastic creating a loose fit with the coaming strips. If the deck cut-out piece has been saved, it should be a perfect fit inside the superstructure sides.

Using the deck access cut-out piece. The superstructure is built over the deck coaming.

The result of this ought to be a superstructure or hatch that securely slides off and on to the deck coamings. This neat fit will, however, have created a narrow gap between coaming and superstructure sides that water can climb up due to capillary action. It is unlikely to result in water entering the hull, but does mean that the inner surfaces of detachable hatches and superstructures need waterproofing to avoid them absorbing water.

Large deck openings can create a potential weak point in a model hull. Perhaps the worst example of this was found in a model based on the small escort carriers built in World War II. For speed and economy of construction, these ships used a simple hull based on commercial practice with the aircraft hangar and flight deck placed on top of what was originally the main deck of the vessel. This looked like a good idea to duplicate and allowed the more detailed and vulnerable parts of the model to be safely out of the way whenever internal access was required. However,

this created what appeared to be a gaping hole in the model's deck and to ensure that the hull's strength was not compromised, substantial coaming strips were made from lite ply.

A similar situation occurred when building the seaplane tender in that an extra deck had been added to these vessels aft of the original amidships' superstructure block. This was to create more internal space for workshops, stores, accommodation and to produce a large, open deck where the aircraft could be worked on. The best model solution was to make this structure lift off a deck coaming in one piece to give uninterrupted access to most of items inside the hull.

MORE SUPERSTRUCTURE

Many warships have only a limited amount of superstructure fitted above the hull deck, but some, especially older vessels, can have multiple

The large deck opening in this model required lite ply coamings to maintain hull strength.

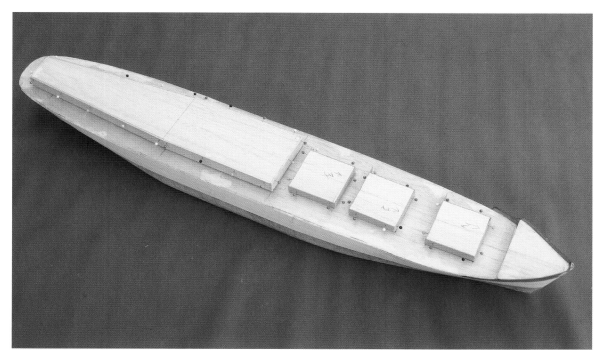

Large access opening allowed the whole superstructure block to lift off the hull.

layers. Before the advent of radar, it was a clear advantage to place things like optical rangefinders as high as possible. In addition, it was sometimes necessary to elevate weapons so that they would have an unobstructed field of fire. A good example of this was the 'Dido' class of light cruisers, which had to mount three twin turrets forward in a superimposed layout, thus adding an extra level to the superstructure.

Probably the best way to build up these superstructures is one layer at a time. This also allows any sections of the decks with overhangs to be added more securely and easily. The block on an aircraft carrier model is a good example of this method. In order to produce smooth airflow over the flight deck, this superstructure

was built with a streamlined section but it also had some cut-outs at different levels, as well as deck projections. Building it up layer by layer, while maintaining a smooth and solid appearance, would have been difficult. The answer was to clamp balsa laminations, with thicknesses that allowed for the decks, and carve/sand them into the smooth shape needed. The laminations were then glued together with the decks inserted between them.

BRIDGES AND BULWARKS

The open bridge type was common on older vessels, where it allowed for a clear view in most

Light cruiser model with triple superimposed turrets.

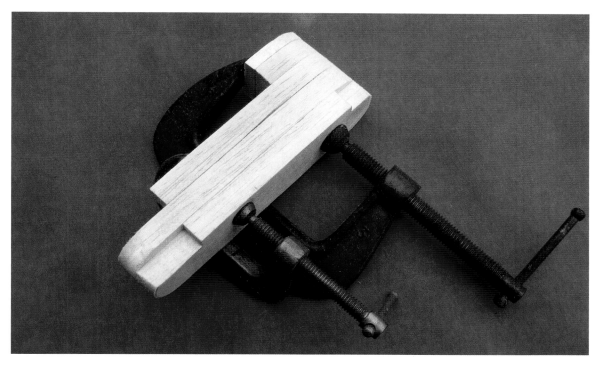

Superstructure laminations clamped together while being shaped.

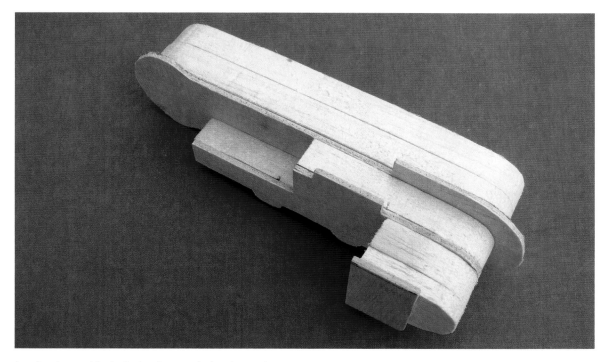

Laminations with decks in place and glued together.

directions – an advantage when both navigating and fighting. To further enhance visibility, the bridge was often placed on top of a superstructure block. This block could be built up from strip and sheet, but it can be practical to shape one up from solid balsa.

Covering the bridge block sides with thin card is also a good idea for creating that neat 'metal' finish. If the bridge is of the open type, then the card sides can be extended above the bridge deck level to create the bulwarks that were around it. Card strips can also be added around various platforms and gun positions including bulwarks. This can call for the card to be bent into a curved shape and, to avoid creases

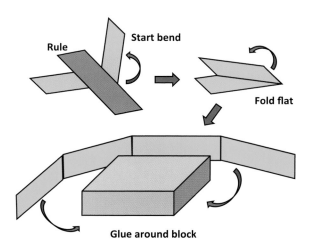

Pre-bending card strips before gluing in place.

Destroyer bridge made from balsa block.

forming, bending around a suitable tube before fitting in place is best. If a platform has sharp corners, then the position can be marked on the card and the corner formed before attaching to the model. Starting the bend using a hard, straight edge (a steel rule is ideal), and finishing it off by folding the card strip flat, can create the desired sharp bulwark corner.

Making and fitting these card bulwarks usually involves a degree of trial and error. It is easy to bend the strip in the wrong direction or length, hindering a correct fit. Luckily, card is cheap and replacing it is a much better option than spoiling the look of the final model. The aim should be a card strip suitably shaped to match its position on the model, so that a light coat of contact adhesive on the mating surfaces will firmly hold it in place. A couple of light coats of the same sealant used for hull and decks, along with very light sanding of the edges if needed, should bond it to the model.

STEALTHY SUPERSTRUCTURES

The Type 23 frigates were designed to minimise radar reflections and had a hull with noticeable flare in the hull sides. This angling was also carried out on superstructure blocks to produce the same effect. Right-angled corners between decks and superstructures would produce a perfect reflector for radar signals.

This model's hull had been designed to use the plan sheet method, so internal access had to be via openings in the deck. Simply building the superstructure over coamings glued perpendicular to the decks would not have created the angled sides that were an obvious feature of these vessels. Angling the coamings was a possibility but would not have had the same secure fitting. Vertical coamings, over which the superstructures slide, are still secure, even with slight upwards movement, whereas angled coamings lose their grip as soon as the superstructure lifts upwards.

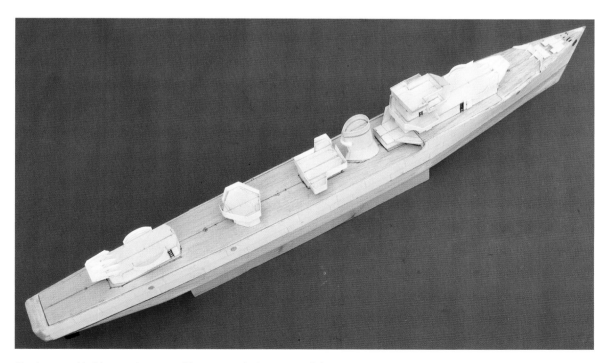

Card around bridge and gun positions on a destroyer model.

Stealthy angled superstructure sides made from trailing edge strips.

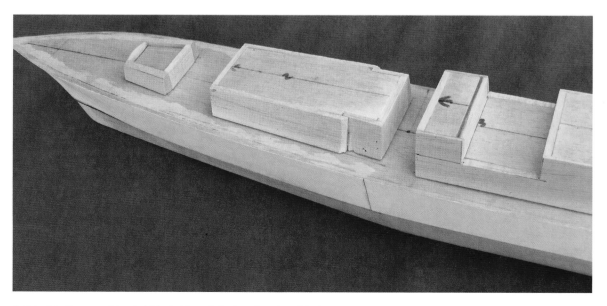

Superstructures cut to right height and deck cut-outs added.

The answer was in an on-line balsawood supplier's stock list. Triangular strips of balsa are usually used to make the rear (trailing) edge of model aircraft wings, so the air separated at the leading edge can join as smoothly as possible after flowing around the wing. Some of these 'trailing edge' (TE) strips were just the right size and shape to fit vertically and securely over deck coming strips, while making the correct angled outer surface.

The TE strips were taller than needed and after gluing together, they could be trimmed to the right height. Because the inner faces of these strips fitted flush with deck coamings, the deck pieces previously cut out could be used for their tops.

The funnel also has to follow these stealthy angles but, being firmly glued to the amidships' block, it was made from balsa sheet. The bridge was of the enclosed type and this was simply made from balsa laminations of the right thickness.

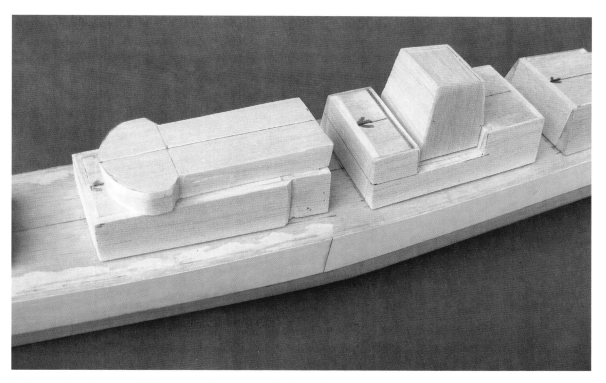

Bridge built up from a solid block, and built-up angled funnel.

Chapter Eleven

Adding Details

After having built the hull and basic superstructure comes a task that can sometimes make or break the model's success. This is the creation of the items that don't actually make the model sail any better but without them, it will never look like a realistic representation of a real vessel. This does not mean that these items have to be perfect and highly detailed, though. However, they must be correctly shaped, positioned and proportioned.

It is worth remembering that looking at a 1/100 scale model at a distance of even 1m (3ft) is the equivalent to looking at the real thing over a distance of 100m (300ft). You cannot expect to see all the small details on the full-size vessel and it gives you a reasonable expectation of what has to be put on the model to create a realistic impression. A good measure can be comparing your model when it is on the water with a similar photograph of the vessel it is based upon.

A possible yardstick to gauge what details can be omitted is to establish a minimum size for details on the model. This could be something like a millimetre; anything that would be below this size could be omitted. Features such as the rigging on masts and railings around the edges of decks can fall into this category. Looking at photographs of the full-size, they can be hard to see. On made to scale sizes they could be delicate and easily damaged items making every sailing session a risky business. Like most things, some degree of discretion is still needed, as some items, like the plating effect on the hull sides, might have to be exaggerated to remain visible in the final model.

It might seem odd, but the addition of a few exquisite (and probably very expensively purchased) items to an otherwise basic model rarely improves the final appearance of the model. If anything, the high quality of these details can actually make the rest of the model look worse. A further problem can be the weight of commercial items made from metal or resin. Added well above the waterline they can reduce the stability of a model. No matter how good such items look, if the model rolls excessively it will at best look silly, at worst make every sailing session an unpleasant experience.

MIXED MATERIALS

Before starting on these items, there is something to be said about the materials to use. Plastic, in the form of sheets of various thicknesses, has a lot to commend it. In fact, plastic in the form of strips, rods, tubes and various shapes and sizes is widely used in the plastic kit and railroad modelling worlds and can make our lives much easier. Plastics do not have any porous surfaces to be sealed before painting and the use of the correct adhesives should produce strong joints. A search for shops or websites catering for these hobbies is worthwhile. They can also be a good place for things like brass or steel wire in sizes that are ideal for many applications in our models.

Suitable materials can often be found around the home. Paper clips are easily straightened and then can be reshaped to fashion things to be added to a model. Drawing pins may have just the right size and shape to form a radar antenna or circular hatches. Bollards can be made from suitable nails or pins. A little imagination will usually come up with a solution.

Many modellers also develop the habit of saving items that might have potential uses in this hobby rather than discarding them. It is not unknown for modellers to suddenly look at an everyday item and declare, 'That could make a good searchlight on the next model'. Cheap ballpoint pens are one such item; their plastic bodies can become perfect for some future project. Even the plastic tube that contained the ink can, if cleaned out, have potential uses; at 1/144 scale, they can make good

'depth charges' on a warship model. Be warned though, this hording of what others might regard as 'useless rubbish' can become unpopular in the domestic home – my suggestion is to hide it out of sight.

ADDING SUPERSTRUCTURE DETAILS

Despite the trend for modern 'stealthy warships' to have clean and uncluttered external surfaces, there are always going to be doors, hatches, lockers and such like somewhere. These can be effectively suggested by the addition of simple but correctly sized and positioned shapes. Doors and hatches can be made from card, just thick enough to remain visible after painting, but not blatantly so.

Lockers are usually just simple rectangular shapes, which can be easily made from plastic. Sometimes you can find plastic strips of the right

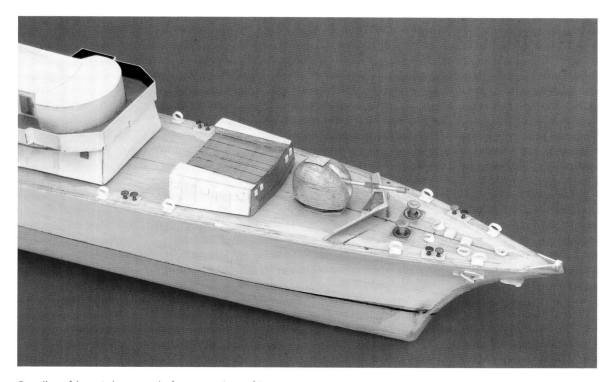

Detail on frigate's bow made from a variety of items.

Small details added to superstructure, funnel and mast.

thickness to cut out the desired shapes. Shops and websites that cater for plastic kit modellers can often supply this material. Failing that, thinner sheets can be glued together to make up the required thickness.

Disks are common features on many ships. Small metal or fibre washers can often be found with the desired diameters and thicknesses. A hole punch, usually used to make holes in paper, can also make handy discs in card and, with a little more effort, plastic sheet.

Ladders are a common feature and I suggest looking in the model railroad area as they can supply such items. You may not be able to exactly match the scale of your model, but a ladder slightly out of scale will probably look much better than a badly built one at the right scale.

If your model has an open bridge, then it ought to have some details added to avoid a very unrealistic empty appearance. Even if you lack details of the bridge layout, similar vessels could give you a good idea of what to install, as the basic items would be similar. This is a small touch that can greatly add to a model's final appearance.

FUNNELS AND MASTS

These can be the first things that strike you when initially viewing a model. One thing to be aware of is that they are not always vertical: a funnel may need to be raked rearwards to ensure that the hot gases from the boilers or engines safely clear away from the vessel. This has often resulted in other

Destroyer's bridge with enough small items to avoid an empty look.

funnels and the masts being given the same rake angle, to create a more visually pleasing profile.

The simplest of funnels can be no more than a tube with a round cross-section. Suitable card tubes may be found inside domestic products. However, these are usually made from spirally wrapped strips of card. No matter how well you seal and paint them, this spiral pattern can stubbornly remain visible. A good idea is to glue a piece of thin card around the tube to create a smooth surface.

Many warships featured a more streamlined funnel shape. Bending an originally round tube into this shape rarely succeeds, as unwanted creases easily form. A better idea is to make a couple of balsa formers that match the desired

funnel section, and fold and glue thin card around them. It is a good idea to recess the upper former a little; this creates a hollow appearance rather than an unrealistic solid one. Also, if the funnel is raked rearwards, the lower former can be angled to match.

Some warships featured funnels with a 'tear drop' section, which can also be made by card wrapped around a former. In this case, the card is creased to create a notch into which the formers are glued, and then the card is glued to the formers. A little gentle pre-bending of the card, around a suitable curved body to avoid creasing, will ease this task. By using an overlength piece of card, after gluing to the notched end, the excess can be trimmed off. I will be honest

A cruiser model with matching rearwards rake of funnels and masts.

Card wrapped around balsa formers to create funnels.

and admit that it can take a couple of goes to get an acceptable funnel, but the expense is minimal and better than a crumpled funnel on the final model.

Funnels are often not left with a plain open top; sometimes a cap and wire guard are fitted. There may be a few smaller pipes visible. Some funnels have a flared base to combine the exhausts from two boilers into a single funnel. This can be omitted and if things like ship's boats hide the base of the funnel, it may not be obvious. If you have to add this feature then some careful 'cut and try' work with a wraparound strip of card is called for. Luckily, card is cheap and it is not too painful to discard any mistakes.

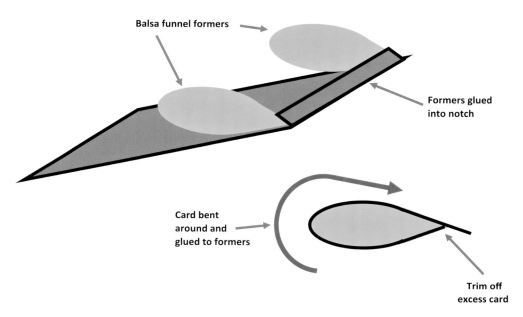

Balsa funnel formers

Formers glued
into notch

Card bent
around and
glued to formers

Trim off
excess card

Making a streamlined funnel with card and balsa.

Extra details added to a card funnel.

MASTS

In some warships, the masts were no more than simple pole types, sufficient to hang signal flags, navigation lights and carry a modest lookout position. In others, a more elaborate tripod or lattice mast had to be used to carry the weight of radio/radar aerials as high as possible. Current warship designs often have the mast as a solid, as opposed to the open tripod/lattice structure, if not a vertical extension of the superstructure.

As what is probably the highest item on a model, masts need to be as light as possible, yet robust enough to cope with any accidental knocks that they can easily be subjected to. For this reason, metal tubes are usually the best material to use. A hollow tube can have more than adequate strength and stiffness yet be much lighter than a solid rod of the same external diameter.

It is often a sensible idea to make masts detachable from the model as this can reduce the chance of damage in storage and transport. A secure fitting can often be made by stepping the bottom ends of the masts into holes in the deck. A simple pole mast might be best fitted into a close-fitting tube fixed into the deck or superstructure. A tripod or lattice mast, with three or four legs, can be secure if pushed into close-fitting holes in the deck.

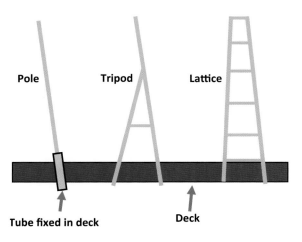

Three styles of mast made from metal tubes.

To ensure that tripod masts are at the correct angle to the decks, it can be best to assemble them '*in situ*' (fitted in place), adjusted to the correct position and then make the joints. Epoxy adhesive has to be used with aluminium tubes but soldering is an option with brass.

Solid masts can be made to be easily detachable yet securely installed by using a plug at their base. I use wood dowels for this, either fixed into the base of the mast and sliding into a matching hole on the deck or vice versa. The same idea was used to make the bridge of an aircraft carrier model detachable, in order to make removal of the deck safer.

Building up a mast, with the various crosspieces, platforms and items they have to carry, can seem daunting. A good approach can be to use a jig to position and support parts while they are joined. If soldering, then drawing the outline on to board and then holding the parts with pins works. The board has to be soft enough to accept the pins but resist the heat of the soldering process – 'softboard' or cork works well.

If epoxy is used rather than solder, then the same technique of pinning the parts to a flat surface can be used. However, to prevent the epoxy from sticking to the board, a sheet of clear plastic has to be placed over the board. This method proved ideal for a lattice mast where soldering one piece in place could easily have unsoldered adjacent joints.

Some mast assemblies are complex and demand a little ingenuity to create the right visual impression. An aircraft carrier was such a model with platforms for radar, navigation aids and other electronic items. These had to be made from plastic sheets and added after soldering the brass parts together. The Type 23 frigate was typical of modern warships with lots of items fixed to the mast. Brass wire, plastic sheet and tubing could suggest many items. The mast also features walkways, which were an obvious feature and could not be omitted but would be a real pain to make. In the end, some ladders, intended for use in the model railway world, were found to be perfect for this job.

Superstructure block secured on two wooden pegs.

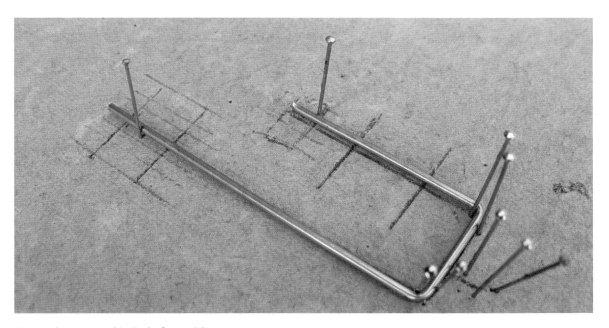

Brass tubes secured in jig before adding next parts.

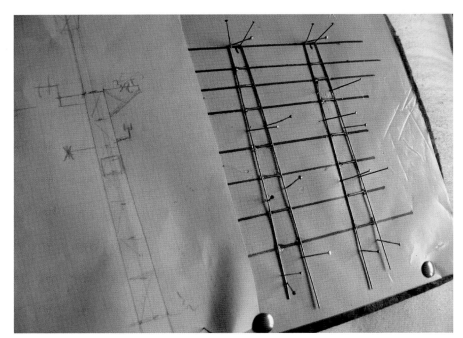

Plan of lattice mast transferred to building jig.

Mast made from brass tubes and plastic-sheet platforms.

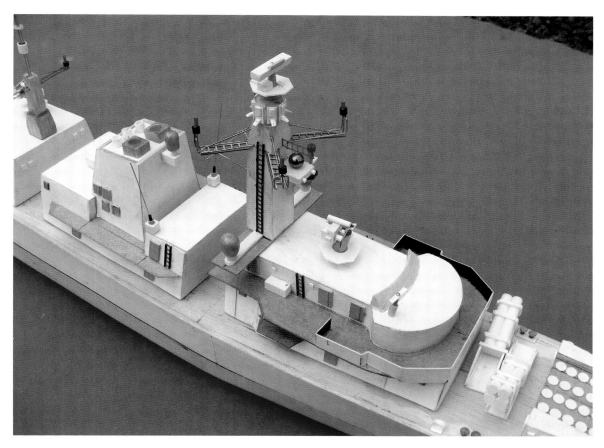

Modern solid type of mast with details from wire, plastic, and so on.

WEAPONS

These are another obvious feature of a warship, although modern designs often build them into the vessel, such as vertically launched missiles that are fired through hatches in the decks. Even these warships are likely to have a few, but still visible, weapons in their profiles.

Guns are likely to be the most obvious weapons needed to outfit a model. They can vary from totally exposed mountings to the fully enclosed type where only the barrel is visible. The exposed type can be a simplified structure of the right size and proportions with just enough detail to suit the scale. A drawing or photographs ought to give a good idea of what to aim for.

Barrels are best made from metal tubes, as plastic ones could bend over time, especially if left in warm conditions. The addition of tubes, rods and plastic can build up the breach mechanism and gun mount. The mount for small weapons could just be a simple shape but some were conical. This is where the ends of some ballpoint pens can find a use. If they are too small then a pencil sharpener applied to some dowel can make a cone shape.

To offer gun crews some protection from the weather and during combat, something more than a simple shield was fitted around the weapon. This could mean avoiding making the breach and mounting gear, but it is usually better to at least suggest these items when the gun is viewed from the rear.

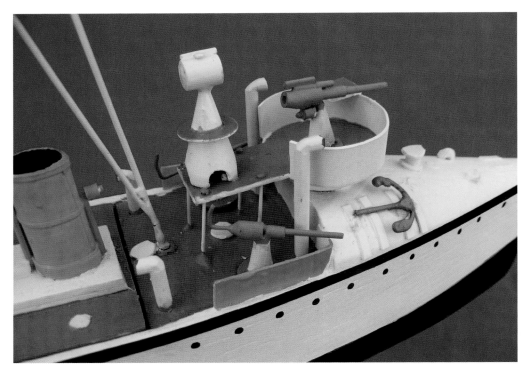

Open gun mountings on an early destroyer.

Twin-gun mounting with open shield at rear.

A vessel that features fully enclosed gun mountings should ease the modeller's task. Some gun houses were of a simple shape, like the US Navy's twin 5-inch mount; however, others were more complex, such as that fitted to the Royal Navy's L and M class destroyers in World War II. They featured a formidable combination of angles, curves and cut-outs that looked daunting at first. The solution was to start with a block of balsa, mark the various angles on it and then start cutting and sanding the unwanted parts away. This model only needed three gun houses, but it is often wise to start making more than you need, being less painful to discard failures as you progress than having to start all over again. If you are left with a few spares, they can always be saved for possible future models

Larger warships had 'turrets' (technically an incorrect term but now widely used), which were fully enclosed and armoured. Some were complex shapes and are perhaps best made by the method previously described. Others had a smoother appearance but with noticeably angled surfaces. To produce an identical set of such turrets, the use of templates to guide their shaping is recommended. The three twin 6-inch turrets needed for a model of the cruiser HMS *Penelope* were made from solid pieces of balsawood. They were first cut to the desired thickness and then had two templates, corresponding to the shape of the turret top and bottom, pinned in place. The balsa outside of the templates was cut and sanded away to produce three identical shapes. After covering with thin card, sealing and painting, they had a solid steel appearance.

MISSILES AND TORPEDOES

In the first missile-armed warships, the missiles were usually stored inside the ship's magazine and loaded onto an external launcher when needed.

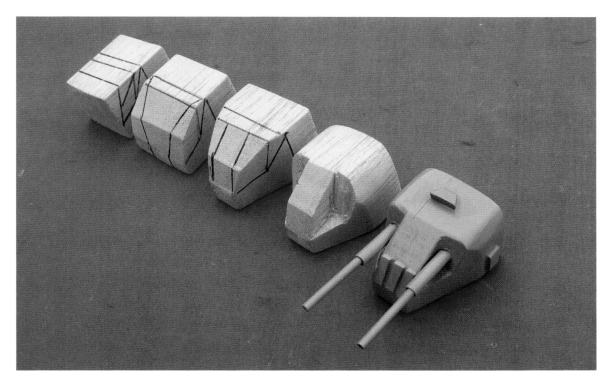

Complex enclosed gun mounting started as a block of balsa.

Armoured turrets made from card-covered balsa shapes.

Missile containers detailed with card/plastic strips.

Modern vessels tend to carry missiles in a 'ready to use' state, either in vertical-launched tubes within the hull, or in external containers. The latter can be made from suitable tubes or rods, and often have minor details such as hatches and ribs around their bodies.

A warship's torpedoes are carried in tubes to protect them and from which to fire them. The tubes are often on trainable mountings, usually fitted so that they can be launched from either side of the vessel. This means that only the heads of the torpedoes would, if at all, be visible. The mounts can vary from carrying a single up to five tubes. Plastic or thin-walled aluminium tubes, along with some the addition of smaller details, can make realistic mountings on a model.

SHIP'S BOATS AND LIFE RAFTS

A warship must carry some smaller craft and these can vary from large launches to smaller oar-driven boats. As with other details, they have to have the size and shape to look correct when placed on the model. The rowing-type of boats can be left open, in which case you have the challenge of building a small model with all the internal details, like planking, seats and oars, being visible. Luckily, to keep the insides of these boats dry, they often have a canvas cover fitted over them.

These small boats were often suspended on davits, which could be swung outwards to launch or recover them. Stiff wire that can be bent into the desired shape can make these davits, which

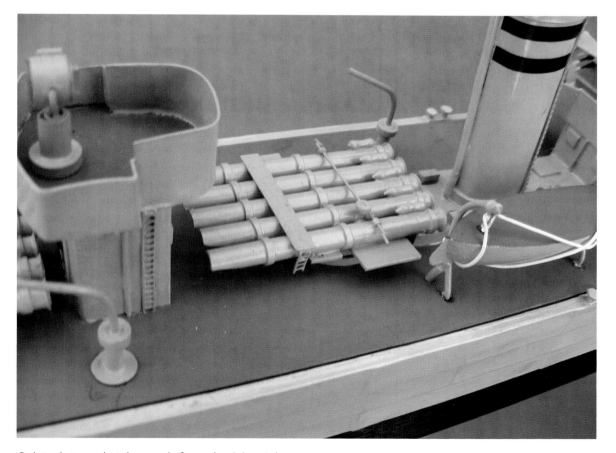

Quintuple torpedo tubes made from aluminium tubes.

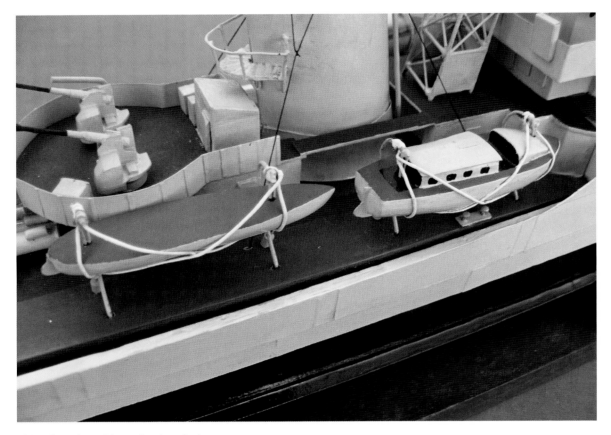

Ship's launch and boat fixed to davits.

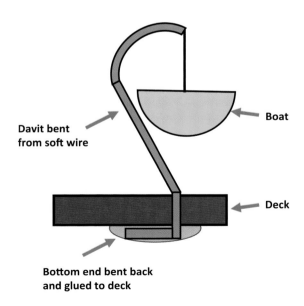

Davit bent
from soft wire

Boat

Deck

Bottom end bent back
and glued to deck

Securing davits into the deck.

can be secured into the deck. Brass wire can be a little unforgiving when bending; copper or even a paperclip (previously straightened of course) can be easier to use. Just gluing them into a hole might not be secure enough, so bending the bottom of the wire back under the deck makes a stronger glued joint.

Larger ships' boats are often placed on deck cradles, launched and recovered by a crane. These cranes may also serve other items, such as any aircraft carried by the warship. If this crane jib has a lattice structure, then it ought to be built in a similar fashion to a lattice mast to avoid an unrealistic, heavy appearance.

Life rafts are common features aboard warships, extra ones often being carried during wartime. Common types have a ring around the edges for buoyancy. The obvious way to make it is to use

Aircraft handling crane with lattice jib made from brass wire.

a flexible tube with a stiff wire core bent to the outline shape. This does not always work as the bent tubes have a tendency to buckle. A perhaps surprising alternative is to make them from solder with the right diameter. Solder can be wrapped around a former, and shaped to match the inside shape of the life rafts without any buckling. The addition of a few weighty life rafts ought not to cause any stability problems on a model not otherwise overloaded with heavy details.

Modern warships may also carry inflatable life rafts in cylindrical containers. They are a much easier item to make, usually from a suitable diameter of rod. These are positioned around the vessel for rapid deployment when needed.

Making life rafts by wrapping solder around a former.

COWL VENTS

These often featured on older vessels, indeed some of the early steam warships were liberally endowed with them. They can be a problem to scratch build, as their distinctive bell-shaped openings must be consistent in appearance. Working in small scales, they can just be 'suggested', but it might be easier to buy suitable commercial plastic vents.

At larger scales, especially if they have to work as ventilation for a steam-powered model, you can make them from a set of kitchen measuring spoons. Cut the end of the spoon free, making a hole in the side, and it can be glued to a tube to duplicate the full-size item. This had to be done on a model based on an early torpedo boat destroyer, as without it the internal steam plant would never have functioned.

Measuring spoons before converting to cowl vents.

Spoons now become working vents on a steam-powered model.

TO CREW OR NOT TO CREW?

Many warship models are sailed without a visible crew, indeed this is probably the norm for modern vessels with an enclosed bridge. However, some types of warship could look quite odd with no signs of life, shades of the *Marie Celeste*?

At small scales, it can be enough to suggest a crew with humanoid shapes. The figures sold for model railroad layouts can often be repurposed to create a vessel's crew. Suitable ready to use figures can be both hard to find and expensive when you do. I suggest bulk-buying mixed packs of the cheaper unpainted figures. It is surprisingly easy to modify them with some careful modelling knife surgery and a little paint. The flight decks of

a few aircraft carrier models have been populated with railway passengers after suitable changes.

With larger scales, figures that are more detailed are needed and military plastic kits can be a good source. A little searching can find kits of soldiers who, with repositioning of limbs, can look they belong on a vessel. There are even a few sets of crew intended to outfit larger plastic warship models that we can commandeer to serve on our models.

One trick to adding figures to a model is to place them in a position where you can view them for a few seconds without it generating an unrealistic 'frozen' impression. On the flight decks, positioned in groups, having pre-flight discussions or preparing the aircraft seems to create the right

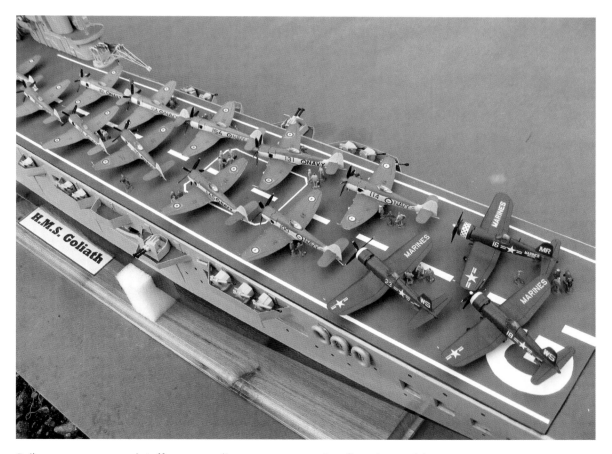

Railway passengers and staff masquerading as crew on an aircraft carrier model.

Military kit items used on a landing craft model.

atmosphere. Having figures engaged in some activity where you can accept a limited movement, if not a motionless state, also works.

AIRCRAFT

A model based on an aircraft carrier would not look right without some visible aircraft on the deck. Likewise, other warships such as battleships and cruisers have often carried flying boats or seaplanes. Nowadays it is hard to find even small warships that cannot operate a helicopter. Therefore, some sort of aircraft is a likely addition to many warship models. At first glance, this might appear to be no problem with the numerous plastic aircraft kits available throughout the

world. The problem is finding ones of the right type to match the model and, perhaps even more difficult, is matching the desired scale.

Probably the most common aircraft kit scale is 1/72 and it is hard to imagine that you would be unable find something suitable if you built in this scale. However, 1/72 is not a common choice for warship models – 1/96 is the nearest one favoured. It might seem that these scales are close enough together to be mixed with no problems. It might be possible, but it is far too easy to create something that will always look awkward.

This can tempt people to use kits of the right scale but wrong type of aircraft. Sometimes this is possible when the kit model has a sufficient similarity with what ought to be used. An example of this was during World War II when the Royal

Helicopter kit secured to frigate's landing pad.

A mixture of US Navy aircraft carried on a small escort carrier.

Navy used versions of the Spitfire and Hurricane aboard its carriers. Kits of these two land-based types abound in many scales and they could be used effectively on our models with the minimum of alterations, if any.

As the design of aircraft has followed common paths, it is possible to find a kit that is of the right scale and layout but a different type. With a little alteration of any obvious mismatching features, careful painting and detailing, it can create the right impression of the correct aircraft. I have done this a few times when having to fill out the flight deck of a model; a few 'rogue' aircraft amongst the correct ones rarely gets spotted, but having too few aircraft does.

There is always the option to 'scratch build' your desired aircraft, which sounds like a challenge. To be honest, building a 'one off' this way is not too hard if you can cannibalise other kits to create it. The seaplane and floatplanes on two cruisers were built this way with wings and tails from one model, fuselage and floats from others. The results were not perfect but had the right scale and looked the part. I would, however, suggest that having to make many such models to a consistent standard would be quite a challenge and it might be better to opt for uniformity and modify kits.

It is worth adding that even if the most thorough Internet search fails to locate any potential aircraft kits, don't stop looking. It is possible to sometimes find the perfect or near perfect item in unexpected places. Toy shelves can occasionally contain some surprisingly realistic aircraft models. I once came across a tray of novelty pencil sharpeners in the shapes of different items; one was a US military helicopter and so was immediately bought and never saw a pencil!

A cruiser's floatplane cannibalised from other plastic kits.

Chapter Twelve

Painting

A good paint job can only enhance the appearance of any model. However, a poor, ill-thought through and possibly rushed job will spoil any model no matter how well it has been built up to that point.

COLOUR CHOICE

This might seem to be simple, just paint the model in the same colours as the full size. This does create the problem of just what colour looks right, especially when viewing a sailing model. As mentioned before on the subject of the amount of small details, looking at our reduced size model is the same as looking at the full size over a considerably larger distance. This means there is a lot of air, which is not perfectly transparent, between the real thing and us. This tends to 'dull' the strength of colours and the effect can be seen everywhere. Take vegetation – close up it might be vivid green, look further away, it is a less strong green, and on the horizon, it becomes a washed-out shade of grey/green. This is something that landscape artists have used for a long time whenever they want to suggest distance.

There is also the problem of different navies having dissimilar choices of colours and these can change over time. There is also the possibility that warships may have to serve in very different conditions. Before the introduction of air-conditioning, vessels serving in tropical regions often had their hulls and superstructures painted white to make the internal conditions better for the crew. There can also be a difference between peace and wartime colour schemes.

Even if the exact colours can be specified, there can inevitably be minor variations; this is especially true if the model is to be depicted in a wartime condition. Add to this the fact that, as soon as a warship has been painted in its official colours, the environment they have to operate in will soon start to change things. In addition, you need to be aware that using photographs to determine the shades can also be a problem in itself. The conditions under which the photographs were taken, like bright sunlight or dull and overcast, can change perceived colours, as can all the processes from camera to final image.

Like most things, the choice of colour has to be something of a compromise. You could try to mix the perfect shade, but this can lead to problems later on. Being realistic and accepting that you might have to touch up a model's paint at some future time, it makes sense to use the best colour match from a readily available range of reliable paints. This has allowed me to repair some hull damage on a model built a couple of decades ago using a new tin of paint that was still a perfect match with the original colour.

Over a century's changes in Royal Navy colours.

GLOSS OR MATT?

It might seem obvious to use matt paints on your model, after all warships are rarely shiny. Plus, matt paints usually have a larger content of colour pigments and so greater covering power. However, matt paints are not as tough as gloss and can be easily scuffed and scratched. Matt paints can be hard to keep clean and they readily show things like the dreaded 'oily fingerprint'.

A compromise can be to use gloss paints on the hull sides and bottom, areas that are most likely to suffer from operating damage. Matt paints can be used on surfaces like the decks and small details that are usually safe from damage. Any visual mismatch this might create can be reduced by giving the final model a light spray with clear varnish. You could use matt varnish, but this could

also suffer from damage like the aforementioned 'oily fingerprint'. For this reason, a clear satin varnish is better: it dulls the gloss down a little, protects the matt and ties the different paints together.

It is worth mentioning that whenever overpainting with different types of paint and varnishes, there is the possibility of adverse chemical reactions. So, when trying a new brand for the first time, it is always wise to have a small trial on some scrap material first.

APPLICATION

There are the two basic ways to paint a model: by brush or spray. Some people are exclusively in one or the other camp, but it can make sense to mix both methods, as appropriate. Both brush and spray painting demand a few common things before they can be successfully carried out.

First, before reaching for any paint, the surfaces to be covered ought to be checked. It is rather too late to discover things like cracks, gaps and pinholes after painting your model. Bare wood must have been sealed, as described in Chapter 3. Applying paint on to unprimed and unsealed surfaces might colour them, but the finish will be horrible. The same comment applies to any porous surfaces, such as card, that you have used in the construction of superstructures and details.

One problem with any paints is that the pigments can settle out over time. Before any paint can be used, it must be thoroughly mixed to restore it to the proper consistency. If the paint has been standing in its tin or jar for any time this mixing must be more than few swirls with a paintbrush. You can get commercial devices to do this, but mine is a small battery-powered hand fan modified to spin a small propeller inside the paint tin/jar. It is powerful enough to mix the paint in a minute or so but without spraying it all over the place.

A single coat of paint may not produce the 'solid' effect that you want. Attempting to achieve this with one thick coat rarely succeeds and usually results in an uneven finish. A first smoothly applied thin coat might look poor and even blotchy, but allowed to dry, and given a second, possibly a third coat, usually gives a much better effect. When recoating a surface it is important to ensure that previous coats of paint have fully dried and, if anything, erring on the generous time between coats. It is also worth remembering that even when the paint is 'touch dry', it can still be soft and easily damaged.

The best painting sequence depends on the model. It is a good idea to think about this before you reach for any paint. Some models could be fully assembled and detailed before painting; others will have a more complex structure that would create many hard to reach surfaces. In most cases, it will be better to paint parts, such as the hull, superstructures and details, before they are assembled together.

THE RIGHT BRUSH

The most expensive brushes in the world will not guarantee a perfect paint finish if they are the wrong size and type for the job. When painting large areas, such as hulls and superstructure blocks, the brushes need to be able to do this with the minimum of strokes yet avoid paint runs from overloaded brushes. A flat brush can be best for this job and a width between 12 and 25mm (½–1in) works in most cases. A steady sweeping action ought to produce a good band of paint on

Homemade paint mixer with small propeller replacing fan.

A collection of paint brushes to cover most areas.

the model. The next band should slightly overlap the previous one but without creating a 'ridge' with it.

An angled brush, like a flat one but with the end cut at an angle rather than square, can be handy when you need to paint up to an edge or in a corner. Round brushes are perhaps best used for smaller items. You will quickly seem to build up a collection, giving you a choice of picking the right size to do the job reliably. Brushes can last a long time provided they are thoroughly cleaned after use. This means something better than a token 'swish' in the appropriate cleaning fluid.

SPRAY PAINTING

You may well own the equipment and have learnt the skills needed to use an airbrush. This can be a very effective way to paint any type of model and many people use it. However, if you have no experience with this method, it is far too complex to describe in a few words and there are many good books that will teach you how to do it.

There is an alternative that might appeal, that is the use of aerosol spray cans of paint. Some of the hobby paint manufacturers offer their product in this form, which can allow you to perfectly match with their tins of paint. Alternatives can be found in DIY stores and auto accessory shops in the form of primers and paints. Care is needed when mixing different types of sprays; it is possible that adverse reactions, such as cracking and crazing, can occur. Hence the suggestion to try spraying on to a test piece before applying to the model. Spray cans of primers are of limited colours but include useful shades such as grey, red, black and white. Primers also have excellent coverage and a couple of light coats can produce a solid colour on the model without hiding any fine details.

Using these cans successfully does require you to follow the instructions. These usually include shaking the can to properly mix the paint and avoiding cold and damp conditions. Before pointing the nozzle at a model, a short blast on some other surface might be wise. This can check that the spray is functioning correctly (no big drops) and the right colour – the latter is a mistake once made, never to be repeated

Spraying should begin with the nozzle pointing off the model, sweeping over it in a smooth

Very useful spray cans of paint and primers.

action and only stopping after leaving the model. This will avoid the excessive build-up of paint and paint runs. It is likely that you will spot some missed areas, especially in the superstructure and on small details. Do not try to immediately rectify them, leave them alone for the recommended time between coats. Note the missed areas and deal with them on the second and any subsequent coats. One of the advantages of using spray cans is the quick drying times and the ability to build up a good finish with several thin coats of paint.

One way to spray smaller items is to stick them to a strip of wood with double sided adhesive tape or tacky putty. The stick can be held at one end while you use the spray can with your other hand.

This allows awkward shapes to be painted by angling the wood strip to ensure complete coverage. It also is more economical on paint compared with spraying each item individually.

DIVIDING LINES

While some items, like smaller details, are often painted one single colour, others will feature two or more colours. This means that a line exists between the colours and usually has to be a sharp and straight one. This can be difficult to achieve by painting freehand. The most troublesome area can be along the hull sides where the colour changes at the waterline.

Small items stuck on a stick before spraying.

Hull above waterline covered with masking tape and paper.

After painting the hull sides from the deck edge to just below the waterline and allowing it to thoroughly dry and harden, masking tape can be stuck with its lower edge matching the waterline. Provided this edge is firmly pressed down, the second colour can be painted up to this edge and slightly over the tape. Carefully peeling the tape away before this paint fully dries can produce the desired straight line. If spraying the second colour, then the previously painted areas need protecting from the inevitable overspray. The common way is to fix sheets of paper to the masking tape as a barrier to the spray.

You can make use of the fact that most vessels have a black band around the hull at the waterline. After painting the hull sides to below the waterline, a black indelible marker pen is used to draw the waterline around the hull, as described in Chapter 9 when preparing the hull for a floatation test. By painting up to this line with a brush loaded with black paint, a neat result can be achieved. An angled brush is good for this task as it makes it easier to blend the paint into the line. If you only 'plate' the hull sides above the waterline then, with care, you can paint the lower hull up to the small ridge it creates.

Now a confession: I usually paint the whole hull of a model below the waterline black. This is contrary to full-size practice, which has a broad, black waterline band, and the rest of the hull painted in some form of anti-fouling compound. This compound is usually a shade of red. Many modellers follow this, but it can lead to an uncomfortable appearance when the model is afloat, especially if a bright red has been used. You can rarely expect to see the whole underside of the full-size vessel when it is afloat, so painting the hull below the waterline black can improve its sailing appearance. It is a personal choice, but if you have to simulate the anti-fouling compound, then a dark shade is best.

PORTHOLES AND WINDOWS

Modern warships may have few, if any, portholes in the hull and superstructures, but many older vessels would look odd without them. If you look at photographs of warships, then the portholes are not a prominent feature and usually just look like dark circles. Some modellers simulate them with eyelets but, even if they create the right size of hole, the rim will stand unrealistically proud of the surface.

The dark-hole effect can be simulated with discs cut from black self-adhesive film. Identical discs can be cut out using a hole punch more usually used to make holes in paper and card. This will only supply discs of one size, which might not match your model's scale. An alternative is to paint them not with a brush, but a rod of suitable diameter. The method is to dip just the square cut end of the rod on to the surface of some gloss black paint. Lifting the rod away from the paint will carry off a drop of paint. To avoid paint runs, the model surface must be horizontal before the next stage. If the rod is placed squarely on the model where you need the porthole and lifted away cleanly, then a black disc will have been left on the model. To get consistent results does take some practice and only one side of the model can be done at a time.

Modern warships tend to have enclosed bridges with windows rather than portholes. These openings could be cut into the bridge structure but would be tricky to do so without weakening it.

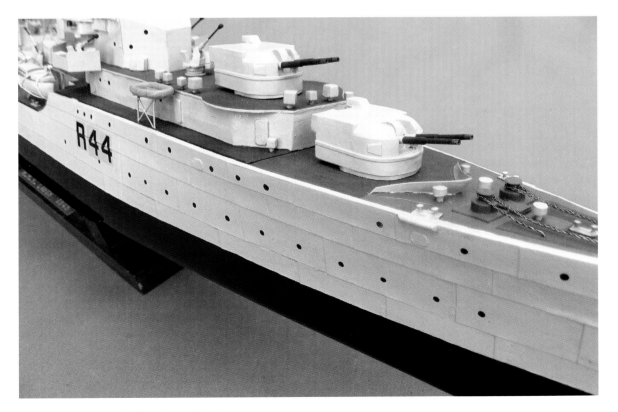

Portholes suggested with blobs of black paint.

Windows on aircraft carrier made from strips of black tape.

It would be difficult to paint these windows and produce the sharp and square appearance needed. Again, black self-adhesive tape can be used. Using one strip of tape, cut to match the depth of the windows and wrap around the bridge; the excess tape between the windows is then cut away. This requires the use of a sharp blade and just enough pressure to cut through the tape. The result is usually worthwhile, even if you have to practise on some scrap first.

LINES, SHAPES, NUMBERS AND LETTERS

Most warships will feature lines somewhere. This can range from warning markings on the decks around weapons that could suddenly move, to lines on flight decks to guide aircraft when landing. You could try painting them, but it demands extreme care and skill to do a good job. The use of masking tape to define the edges of the lines when painting can be used for straight lines.

It is much easier to use some strips of self-adhesive tape, such as pinstripe, as used on automobiles, or cut from sheets of adhesive-backed plastic film, as sold in DIY or hardware stores. This is especially true when making curved lines, such as the circles often seen on flight decks to guide helicopters. In this case, the use of a circle cutter is strongly advised.

Warships often feature some form of identifying letters and numbers on the hull sides. Flight decks can also have something to help pilots

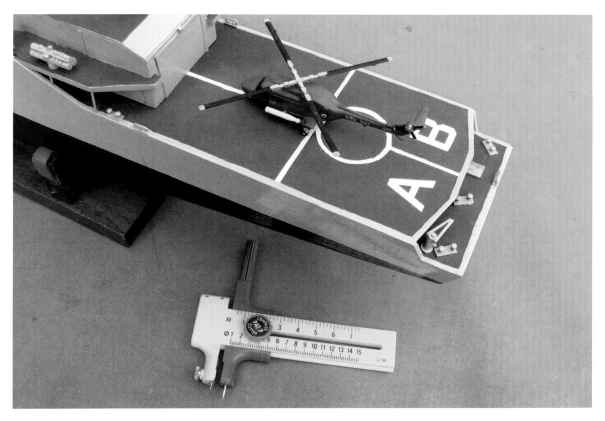

Landing pad markings made from tape using a circle cutter.

land on the correct vessel. If water-slide transfers (decals) are used, then they must be sealed with a clear varnish or your model is likely to return from a sailing session in anonymous mode.

Matching the size and style of letters and numbers needed for a model can be a problem. It is possible to cut your own out of adhesive-backed plastic film, but be prepared for lots of scrap, especially where complex curved patterns are needed. Luckily, an Internet search can often locate suitable ready to use adhesive letters and numbers in a range of styles and sizes to match those of many navies. It is also possible to find something suitable in craft and stationery stores.

It is difficult to paint these letters and numbers neatly on to the hull or deck and match the smart appearance expected on a warship in peacetime. However, if the model is being built to represent

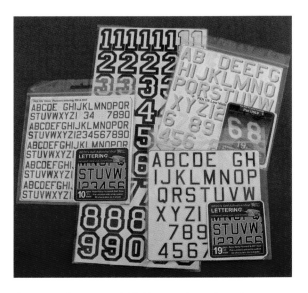

A range of commercial letters and numbers in different sizes and styles.

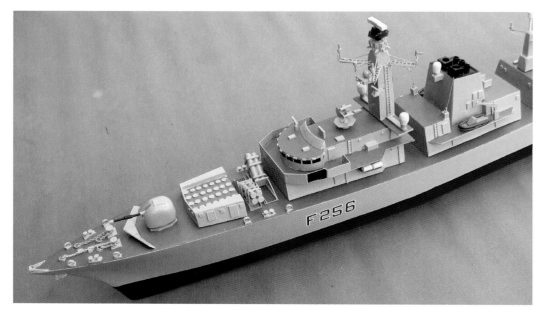

Identification letters and numbers added to model sides.

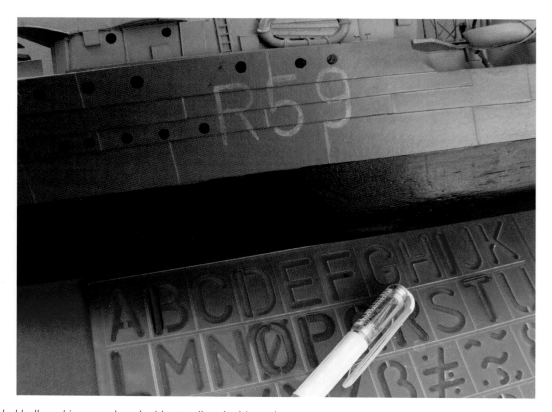

Faded hull markings produced with stencil and white gel pen.

how it would look in wartime, then a less than perfect job might be appropriate. Rather than using paint, it is possible to get a white gel pen from art supply shops. This, when used with a suitable stencil, can produce consistent letters/numbers on a model. The strength of the white colour can be controlled by the number of times you draw the pen over the surface. This proved ideal on a model destroyer where I wanted to reproduce a still visible, but heavily faded, hull identification. The gel is not waterproof and needs protecting with a thin coat of clear varnish.

STICKING BACK TOGETHER

After all the detail items have been painted, they can be secured to the model, but first, a 'dry run' might be wise. This is no more than placing the items on the model, standing back and looking

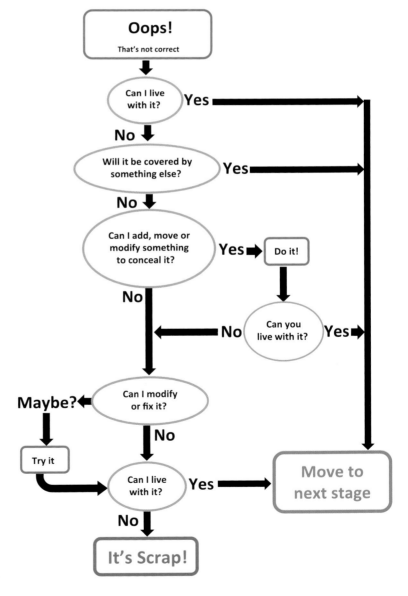

How to solve problems with details.

at it critically, possibly comparing it with photographs or plans of the real vessels. It is much better to do this rather than finding out that items that looked correct individually, just do not work when placed on the model but are now well and truly stuck in place.

If something is a problem, then do not despair – this decision tree shown on the right (courtesy of Roy Cheers and Peter Fulcher) may be able to get you out of trouble. If you have made more items than you need for the model, always a good idea where multiple identical items are needed, it can sometimes be possible to replace it with a better-looking one. Even if this is not possible and you have to make a new item, save the old one as it might find a use in another model.

If these detail items have been made as light as possible to ensure a stable model, then it is possible to secure them with equally light methods. A little glue of the domestic contact/impact type can be more than adequate. Latex glues have the advantage of making a secure bond but with the ability to be peeled apart with minimal risk of damage later. I have even used double-sided adhesive tape where there were two suitable surfaces to join. Small balls of the tacky putty, used to hold paper on walls, can secure items surprisingly well. It can also allow items to be safely removed with a steady pull. 'Superglues' can be used, but if they work as advertised, you have to place things exactly right with no chance to make adjustments.

Aircraft and helicopters are things that seem to attract accidental knocks. If rigidly fixed to the model, this can easily lead to breaking delicate items off them – propellers and rotor blades especially. To allow a little 'give' in these items,

Helicopter secured to landing pad with two wires.

they can be built with wires, about 1mm in diameter, protruding from their undersides. The wires should be long enough to fit through holes in the deck. This alone can keep them in place when the model is sailing but, if needed, a small amount of tacky putty can be used where the wire exits the deck's underside.

WEATHERING?

The commanders of warships are usually fastidious about the appearance of their vessels and regularly have their crew touching up any less than perfect paintwork. This means that we modellers can usually justify finishing our creations with a smart blemish-free appearance.

The situation can be different if you want to create the image of a vessel in wartime or even after other arduous service. Since most of the full-size vessel's structure will have been made from steel, this means 'rust' will be visible. Checking photographs of the real things can give you a good idea of what it could look like. Rust can occur anywhere, but damage, corners and edges are usually favoured starting points.

It is too easy to overdo weathering and slap an excessive amount of rust-coloured paint over a model, which never looks right. It is much better to build up the effect with greatly thinned coats of paint. Heavily rusted areas can be highlighted by using the 'dry brush' technique, by using a paint brush on which most of the rust paint has been wiped off and gently brushing it over the area where rust is required. Repeated application can build up the desired rust-stained effect. Likewise, a light wash of dirty thinners/cleaning fluid can add to the grubby appearance.

The key to successful weathering of any model is to practise on some scrap material and learn when to stop just before you overdo the effect. If a mistake is made on the model, you might have to repaint it or concoct a reason why the vessel's commanding officer has let it get that bad.

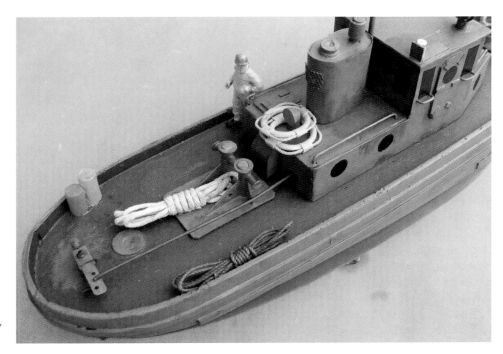

Just enough weathering on a small tug used by the US Army.

Chapter Thirteen

Preparing to Sail

I must confess that many of my models started their sailing lives with some of the smaller details still to be added. This allowed me to check the model's sailing performance and correct any problems as soon as possible. To be honest, there could also be some impatience on my part with the strong desire to see my efforts finally working.

An early floatation trial when the hull had been built was suggested in Chapter 9. This was a quick test to check the internal layout and operation of motor and RC system, plus the model's stability. The rudder servo and motor had been properly installed, but the ballast, battery and other items only loosely fitted. It would be unwise to repeat this temporary installation and dash off on the maiden voyage. You want nothing to fail when the model is out of reach.

RE-BALLASTING TRIALS

This is best carried out in safe and calm water – the bath is ideal if the domestic situation allows. The experience of a previous floatation trial will have established the approximate positions of things, but the addition of decks, superstructures and details will have probably altered the trim a little. The first thing to do is to get the model floating upright and on the desired waterline.

The battery, receiver and ESCs are the most likely items that you will need to remove regularly and so can be held in place with foam plastic. The receiver and ESC, being light, can be held securely in slightly undersize cut-outs made in foam blocks that are themselves wedged in to the hull compartment. Batteries can also be secured this way but the heavier lead-acid types might be better situated with a wooden frame. The test is that these items can be installed and removed safely without the risk of excessive effort being required.

The ballast must be as low as possible inside the hull. Some degree of moving internal items and adding/removing ballast is almost inevitable before the right trim is achieved. After refitting the decks and superstructure, the crucial transverse stability needs checking with a push down on one side until the deck edge is at the waterline. Release the model and, if it smartly rolls back upright, usually with a few decreasing oscillations before coming to rest, then you should be safe. If it fails to do this, then think about how to lower the model's centre of gravity. For example, moving internal items lower and checking that the upperworks and details are as light as they possibly can be. Adding extra ballast, again as low as possible, is another option. It may make the model sit a little lower, but this is not always obvious and much better than a model that is never comfortable to sail. There is the option of adding external ballast under the hull, which, if painted black, is hard to see when sailing, and, if easily detachable, it does not spoil the looks out of the water.

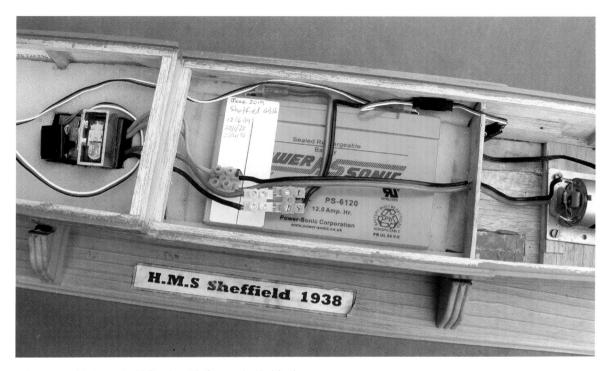

RC gear and battery held firmly with foam plastic blocks.

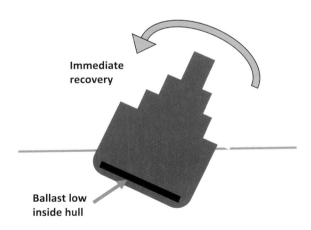

Immediate recovery

Ballast low inside hull

Testing transverse stability before sailing the model.

When satisfied, it is important to make sure that no ballast can move, even if it seems firmly wedged in place. The odd knock or shake when handling or transporting the model can move ballast. It can also move when sailing due to the action of wind, waves and even other modellers. The result can vary from a comical list while you recover the model, to something much worse. For this reason, it is a good idea to 'tack' ballast in place. This only requires a few spots of glue between the edges of the ballast pieces and the hull. To make it possible to remove the ballast without damaging the model, I often use silicone sealant for this job, as it bonds well but can be peeled or cut away safely.

RADIO CHECK

With the model properly outfitted and ballasted, it is sensible to give the motor and radio control gear a precautionary workout. Before this, all the wiring and connections need sorting out into a practical layout. This is one that allows easy access to things like the battery/ESC plugs and sockets that need unplugging for recharging and when

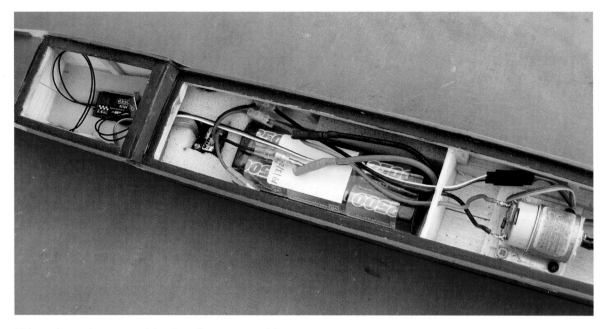

Tidy and easy to access wiring in a destroyer model.

not using the model. The receiver/servo/ESC connections are usually left alone unless something needs replacing, but if the ESC has a BEC (which uses the drive battery to power the radio system) there should be an on/off switch that needs to be in a convenient place. Even with the RC system switched off, an ESC connected to the battery can still be drawing a small current and, left for a long time, can flatten and possibly ruin the battery.

You could cut the wires and cables to the exact length for installation in the model, but this might cause problems if you decide to change things or transfer items to another model. To avoid the risk of overlong wires becoming entangled with anything, it has always seemed better to just loosely coil up the excess and push it safely out of the way; between any foam packing and the hull sides is often a convenient place. You do not need to make a wiring installation to the highest professional standards, but you want one that is tidy and safe to use, especially if a little 'trouble shooting' should ever be needed. You definitely

do not want a mare's nest of multi-coloured wires that would be more at home in a modern art gallery rather than a functional model.

TRANSPORT AND STORAGE BOX

While the model may be robust enough to avoid any minor accidents, some protection during transport to and from the sailing water is a worthwhile addition. It can also make a safe storage place for the model when not in use. I will confess that this is when my models are most likely to suffer accidents.

This need be no more than a simple box into which the model can be placed along with a little packing to prevent any movement. Cardboard boxes can be used for smaller models, but they often need some reshaping to suit the model's size. Unless the outer surfaces are painted, bare cardboard will absorb water and lose its strength – embarrassing if caught in the rain during a sailing session.

It is not too difficult, especially after building your model, to make a wooden storage and transport box for the model. A simple rectangular box, sized to match the model, can be made using wood for the top, bottom and two end pieces. Having detachable masts can help to keep these boxes to a more convenient size.

The box sides can be made from plywood or hardboard, the back being pinned and glued in place. For safe placement and removal of the model, the front panel can be hinged at the bottom with catches on the top edge. A handle secured to the middle of the top piece makes it easy to lift the box and model. Foam packing at the bows, stern and sides stops any movement and potential damage to the model. One tip about these boxes is not to over-build them; they need to be just strong enough to protect the model from accidental knocks and bumps. This was brought home to me when I built a box from some pieces of lumber just because they were handy. This resulted in me having to carry four times the model's weight between the car park and sailing water.

TOOLS AND STUFF TO TAKE

It is something of a truism that whichever tool you leave at home, it is going to be the one that you need to keep you sailing. There should be no screw, nut, bolt or whatever in the model for which you do not have the appropriate tool when at the sailing water.

Cruiser model safe in its box. Note masts detached and stored in the box.

As many connectors in rudder linkages and motor couplings need Allen keys to secure them, they must be taken. In fact, since these keys can be supplied with the connector/coupling, it is not a bad idea to put them somewhere inside the model. If a problem arises then they are immediately to hand.

One 'tool' that can be a godsend should any electrical problems arise, is a small multimeter. Checking battery voltages is an obvious task for these meters but they can locate troublesome 'breaks' in electrical circuits that are otherwise invisible.

Pliers of the needle-nose type are very handy for holding small items and picking up things that have fallen inside the hull. It is not really recommended, but these pliers can be used in place of spanners when a nut has to be tightened and you do not have a spanner.

Safely launching and recovering the model is something to consider. Some small models can be simply hand-launched and recovered, but weightier ones could be troublesome with the risk of damage to the model and even the modeller falling into the water! Some elaborate launching aids have been made, but a simple one is two loops of cord with a handle. The handle can be made from wood; dowel (something like a broom handle) is ideal and should allow the cord to lift the model without risking damage to it. The cord length should allow the model to be comfortably transported to and from the water with one loop under the bows and the other under the stern. Provided the model is kept level, there is little risk of it sliding out of the loops.

Although not 'tools' as such, spare propellers are handy things. Propellers can become damaged and even drop off if not secured firmly. In addition, when starting with a new model, it is handy to take a variety of different propeller types and sizes to check which gives the best performance.

Also not tools, but useful items can be tissue, of the kitchen roll kind, and some antiseptic hand gel or wipes. The tissue allows you to dry the external hull surfaces after sailing and before replacing the model in its transport box. Should any water have entered the hull, the tissue can mop it up, perhaps as a wad held in the jaws of the pliers. Since the sailing water is unlikely to be

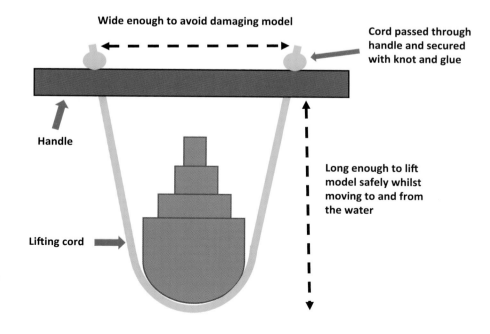

Wide enough to avoid damaging model

Cord passed through handle and secured with knot and glue

Handle

Long enough to lift model safely whilst moving to and from the water

Lifting cord

Lifting straps for safe model launch and recovery.

Just about everything to keep me sailing is in this case.

pure, it is a good habit to clean your hands with antiseptic gel after sailing; the risk may be small but why take it?

If your RC outfit uses dry (non-rechargeable) batteries, then taking a pack of spares is a wise thing. Nothing spoils your day like seeing the transmitter's battery meter fall into the danger reading after the first few minutes of sailing.

You may find that the model's trim, although it looked perfect during ballasting trials, is not quite right. Taking some small pieces of ballast material can be useful for any final adjustments that are needed. I usually have a few bits of lead for this job but have been known to use some small coins.

Rather than stuff your pockets with these items, they can be packed into a suitable case or bag. One of those metal tool cases can hold just about everything. Mine is set up for serious sailing sessions with two transmitters and just about all the items I might need short of a total model rebuild. Using cut-out's in foam plastic, everything is held firmly and safely. It also carries the instruction manuals for RC gear; some outfits have a quite complex system for making adjustments and different makes have their own idiosyncrasies. A manual left at home is little use when you need its contents at the lakeside.

BEFORE SETTING OFF

A checklist is a good idea and widely used in jobs where a small mistake or omission can have serious repercussions. Are the batteries freshly and fully charged? It is not unknown for a modeller looking at their stranded model in the middle of a lake to cry, 'But I only charged the battery six months ago'.

This applies to both the model's drive battery and, if using a rechargeable battery, the RC gear.

Has the model's operation been tested before setting off? This need only be a spot of lubrication on the motor and propeller shaft bearings, connecting the drive battery, switching the RC gear on and checking that everything functions correctly. It is also important that you remember to switch off and disconnect the battery, especially if you carry out this test the night before you go sailing.

With a properly sorted and reliable model, the weather need not be perfect to allow you to sail your model. However, if it is new or your first model, then a maiden voyage in rough and blustery conditions will not tell you much about its sailing characteristics. It does not need to be flat calm, but any wind should be light and steady.

Chapter Fourteen

Sailing

The first, and sometimes forgotten, requirement for sailing any model boat is safe water in which to operate. Safety refers to both you and the model. Having to negotiate a slippery bank to launch and recover your creation is never a wise thing to attempt. Likewise, having to wade out into water deep enough to float the model is rarely a good idea.

Just in case of problems, it makes sense to operate on the down-wind side of the water, if possible. Should the model unexpectedly stop, then the wind will blow the model back to

A model's first sail starting off with the correct static trim.

you. This can be safer than allowing any fellow modellers to attempt to push it back to you. Some want-to-be salvors can be dangerously over-enthusiastic in their actions! A check of the RC functions before launching, even if you did one before leaving home, is sensible. It has been known for people to launch a model without switching the RC system on.

After placing the model in the water, its trim should be noted; it is possible for things to move a little while being transported to the sailing water. If correct, with the model pointed away from the bankside, the motor stick can be advanced slowly, just enough to start the model moving away from you. This is another precaution, just in case the motor has been incorrectly connected and it starts to move backwards. Another precaution is to test the steering at this stage and check that a right rudder command at the transmitter does actually make the model turn to the right.

With a little more power applied, the steering response can be checked out. The first thing to note is, does the model run straight with the transmitter's rudder stick or wheel at the centre position? More than likely a slight turn will be noted and can be corrected with the rudder trim. Do not confuse this with the drift that any crosswinds can create, though. If this turn is pronounced, it suggests that the rudder–servo linkage is at fault. You will need to check the rudder's response to steering commands: how quickly and tightly it turns to different movements of the transmitter controls.

RUDDER PROBLEMS

For general sailing, a maximum rudder angle of around 30 degrees either side of the neutral position is a good starting point. This ought to give a smooth rudder response and a safe turning circle with a diameter of around four model lengths. If you feel that the model has a poor turning circle, then the rudder throw can be increased by either moving the link on the tiller arm to a hole closer to

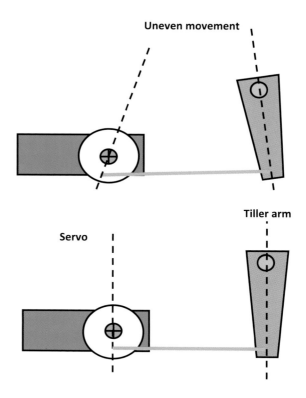

Square rudder linkages for even rudder moment.

the rudder shaft or further out on the servo arm. If you do this, check that the steering system can still move freely with no risk of becoming jammed at any position. If the model turns too tightly, just reverse these actions. Some transmitters have the ability to alter the maximum rotation of a servo; this could be used as an alternative to altering the rudder linkage. The RC outfit's instructions ought to cover how to do this.

Another steering problem can be uneven turns, which is noticeably tighter in one direction than the other. The usual cause of this is not having the servo and tiller arms square. This means that, even though the servo's movement in either direction is equal, the tiller's movements are not. Adjusting the linkage is the solution.

Gradually increasing the power will allow you to test the model's steering at higher speeds. You might notice that tight turns cause the model to

Model safely healing to the outside of the circle.

slow down. This is due to the model no longer sailing smoothly through the water but moving at an angle with the bows pointing into the circle and creating more resistance. The model may heel to the outside of the circle when turning at speed. This is normal and the full-size vessels usually do this. With adequate stability, this ought never to be a problem and on returning to a straight course, the model should immediately roll upright. Overpowered models, along with excessively large rudders, can cause a model to heel to the inside of the circle, but this can also be safe, if somewhat unrealistic, behaviour.

WANDERING MODELS

You can sometimes experience a model that will not return to a straight course after centring the rudder stick or wheel on the transmitter. This can be due to 'looseness' in the linkage between the rudder and tiller arms. My test for this is to put a finger on the rear of the rudder blade and gently see if I can move it. What I look for is a side-to-side movement that I can barely feel but hardly see. Any more than this and you have to look for things like overlarge holes in the arms, flexible connections and, possibly, a worn-out servo, where its output shaft has noticeable free movement.

There is also the opposite cause with a very 'tight' linkage, where the rudder servo is struggling to move the servo back to the neutral position. If the servo moves noticeably slower when connected to the linkage or just sounds like it is struggling (usual sign is a buzzing noise), then these are common warning signs.

There is another possibility. The hulls of high-speed warships, such as destroyers and cruisers, are slim and usually have good, straight running characteristics. Those with fuller hull forms, like the escort carriers built on merchant ship hulls during World War II, have quite bluff bows that can hinder returning to a straight course. The solution, when straightening out of a turn, is to apply a touch of opposite rudder until the model's heading is correct. Once appreciated, this can become an automatic action when sailing.

Bluff bows of this model need small rudder corrections to maintain sailing course.

SAILING BACKWARDS

While most propellers are designed to work most efficiently when driving the model forwards, if run in the opposite direction the model should still sail astern. The maximum astern speed will be less and, to be honest, sailing full-power backwards can cause problems. The stern of the model is usually blunter than the bows and water can build up and even flood over the deck. The extra resistance of the water pushing against the stern can also make the model impossible to control; it can enter into a turn that no amount of rudder movement can correct.

Some models with true-scale hull forms can never be fully controlled when sailing astern, but the simplified shapes used in SOS models are usually much better behaved. The trick is to find the comfortable astern speed that is just fast enough for the rudder to have control but not too fast for any wayward tendencies to overpower it.

SLOWING DOWN

Model boats, just like full-size vessels, do not immediately stop as soon as you cut the motor power. Warships have quite sleek hull shapes and can glide on for some distance before coming to rest. It is a good idea to find out what the stopping distance from full speed is. A handy way to think about it is in terms of model lengths, for example, something like four or five model lengths.

A good habit to develop when sailing is to use this stopping distance to create a danger zone

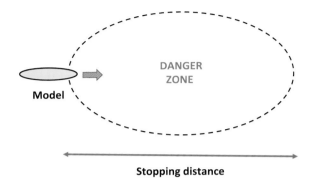

Stopping distance

Using the model's stopping distance to create a danger zone.

ahead of the model. When sailing and keeping your eyes on both the model and this danger zone, you can take immediate precautionary action if anything enters this zone. It is much safer not to assume that other boaters will do the same.

You might think that full-astern power will always get you out of trouble, but this can cause the propeller to aerate or ventilate. This is when the propeller, especially when close to the water surface, can suck down air and create a frothy air/water mixture around itself. In this condition, the propeller creates little thrust. An alternative action when faced with danger is to cut the power and apply full rudder. This slews the model into a turn, which immediately slows the model and can turn it away from danger. If it is possible to safely turn away, then keeping the motor powered can also work.

RIGHT SPEED

The subject of suitable speeds was discussed in Chapter 2. This can be checked in these sailing trials. It need not be a difficult job but good enough to show if we are in the 'right ballpark'. I simply time the model as it sails past two markers on the bankside a known distance apart. This is good enough to show if the model is slower or faster that it ought to be. A small change to the propeller size (*see* Chapter 4) can easily alter the top speed.

AFTER-SAILING CHECKS

Immediately on removing the model from the water, it is a sensible idea to open it up and inspect the internals. Does the RC system still function correctly, are linkages secure and has anything become loose or moved? The worst thing is perhaps finding some water inside the hull; its location must be noted and any free water immediately removed – this is where some kitchen roll tissue is handy. Touch a fingertip on the motor body and if it is warm, that is fine, but if it is uncomfortably hot, it needs investigating.

Unless you are very lucky, the first sailing session of a new model will introduce a few problems. These might only be minor, like trim adjustments or the model turning better one way than the other. Such problems need investigating back at home. It is easy to forget things or even correctly remembering which way the model was listing. Making up a 'gripe sheet', listing the problems as they occur when sailing, is a way to avoid this difficulty.

WET MODELS

Even if the insides of my models appear dry after sailing, I still leave them opened up with hatches and decks removed for a day or two. This might seem to be unnecessary, but even an unnoticed drop or two of water left inside a sealed hull can lead to corrosion or decay. Having always done this means that despite many of the models illustrated in this book being decades old, they are all still sound and operational.

If significant water has entered the hull, all electronic items, such as the receiver, servo and ESC, must be removed from the model. If these items have been immersed in water, then the ideal thing would be to open up them up and place them somewhere warm to dry out. This is not always easy as some are difficult, if not impossible, to open without damage and invalidating any guarantees.

Luckily, ESCs are often totally sealed and water-proof so just need to be dried. Servos are usually quite well sealed and, unless fully submerged, can resist water ingression. Receivers' cases have openings for servo plugs, so water can readily enter. However, if promptly dried, damage can be avoided. A critical check of these items when rein-stalled in the model is necessary. If their operation is in any way suspect, replacement is the wisest action.

Batteries can better resist water; sealed lead-acid types are exactly that, nothing can get in or out. Other types can be sealed in plastic, which is unlikely to be totally waterproof, and cutting open the cover is not a good idea. The best answer is leaving in a warm, dry place and checking that they behave properly when recharged.

The other things to check are any switches and connections in the electrical circuits. If left wet they can suffer corrosion that can lead to increased electrical resistance, which is never a good idea in a model's power supply. Again, if in anyway suspect, replacement is recommended.

One final point: these comments only apply if you sail in freshwater. Sailing in saltwater is a very different situation. Metals can corrode rapidly if in contact with saltwater. Only models especially prepared for these conditions and correctly treated post-sailing should be allowed in saltwater.

LEAKING MODELS

It is vital to find out how water entered the hull. Leaking through deck and hatch openings should not be a problem unless you sail in conditions where water washes over the model. The answer here is simply not to sail in such conditions again.

No matter how carefully you sail your creation, checking for hull damage after every sailing session is wise. Hulls made from wood can be surprisingly tough provided all the joints were properly glued together and the surfaces sealed and painted. Punching a hole in the hull takes quite some effort and is unlikely to go unnoticed; however, scrapes and scratches can be missed. They may not cause a leak but could let water soak into the wood. After giving any damp wood a chance to dry out, the damage must be sealed and repainted.

A likely location of leaks is to be where the propel-ler and rudder tubes pass through the bottom of the hull. Any 'pinholes' or hairline cracks in the glued joints between hull and tubes can be hard to locate. After the hull has fully dried you could run some epoxy around these joints, slightly roughening the surfaces first for better adhesion. Repainting this area can also help to seal minor defects.

It is possible that water can enter the hull via the propeller and rudder tubes. The top of the rudder tube will probably be above the water level when sailing, but the inner end of the propeller tube is likely to be below this level. If the propeller and rudder shafts are a good fit in their tubes and they have been installed correctly, then they should give no problems. It is worth checking that there is no excessive movement between the propeller shaft and the tube bearings. Also, that the lock (jam) nut or washer is pressing on the face of the bottom bearing when the propeller is pushing the model ahead. A few drops of water from the inner end of the propeller tube after a sailing session is quite common and not something to worry about, but a motor compartment full of water is.

DRIVE LINE PROBLEMS

It was suggested that you give the motor a touch test after sailing. If warm, that is usually no problem, but if hot, something is wrong. Before investigating such problems, for safety the motor must be disconnected from the battery. One check for smooth transfer of power from motor to the propeller is to gently rotate the propeller with your fingertips but with the power *off*. You will feel the poles of the motor align in the magnetic field, a slight resistance as you rotate them out of alignment, then they jump to the next position. If this cannot be felt, then there is some unwanted

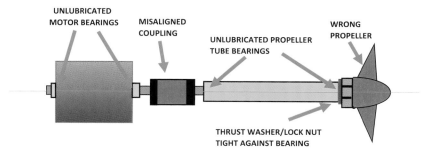

Possible problems in a model's drive line.

friction somewhere. Disconnecting the coupling can help to locate it.

Another possible problem could be the wrong propeller. Changing to a slightly smaller one could allow the motor to operate with greater efficiency and so create less waste heat. As the motor would be spinning the propeller faster, the model's speed may not be changed. In addition, a different propeller with the same diameter but less pitch can make a great difference. It pays to check and be prepared to experiment in the early days of sailing your models.

With care, you ought to start sailing with every confidence that the model will come back to you.

Chapter Fifteen

What Next?

Having successfully designed, built and got your model reliably working to the state where you can take it to the sailing water, switch on and sail with confidence, you might start to think about another model. This is a common occurrence as it is human nature to feel that while this one has been a satisfying experience, although maybe with a few problems that needed solving along the way, the next one could be better. There is also the possible challenge of trying new ideas with the next model.

It is hard to build a warship model without learning something about the reasons for its existence and the history of its service. This can often lead people to building models based on a class of vessels or those serving in particular navies. I will confess to having a fondness to warship models from the World War II period onwards. Royal Navy frigates, destroyers and cruisers have been favourites, but vessels of different types, periods and navies have been too tempting not to build.

A list of useful authors is included in the Resources, along with some of their work that I have found to be extremely helpful. It is by no means a fully comprehensive list, but using these names and a search in a library or booksellers' Internet listings can locate many more possible reference items. These types of books often have a limited time in print and so can be rare and expensive; hence, borrowing from the public library system is a good option.

It is sensible to start with a model where only the speed and direction are controlled by a radio link.

Once mastered, you can think about adding extra features such as lights (for night-time sailings), sound and smoke effects. There is, however, one thing to be aware of and that is potential stability problems if these extras are weighty and placed high in the model. Luckily, modern commercial electronic items are often small, light and reliable.

It is often recommended for newcomers to this hobby to join their local model boat club. This can be a fount of useful help, advice and encouragement, but only if there is one nearby. An Internet search can locate these clubs, and their websites ought to give clear information about the location, facilities and activities that they can offer you. There are also other websites that cater for this hobby: in the UK there is 'Model Boat Mayhem' and one run by the magazine *Model Boats*; the American 'RCGroups' website also has a popular section for model boats. Many other sites exist, but some care needs to be exercised as to the relevance and value of any advice offered through this medium.

A final point in favour of this and similar hobbies is the skills it develops, not only in use of materials and tools. Creating an original working model demands planning and anticipation, and the almost inevitable problem-solving experience. These talents may be learnt on a modest scale when building a model but can be vital when faced with problems that are larger and more serious. However, even without this, it can simply be an absorbing, relaxing and enjoyable pastime.

Resources

BIBLIOGRAPHY

The following list is by no means comprehensive, but provides a useful list of authors and some of their works. Using these names and book titles in an on-line library or booksellers' search systems should allow you to locate many, many more books that have potential value in this hobby.

Chesneau, Roger *Aircraft Carriers* (9781860198755)
Friedman, Norman *British Destroyers & Frigates* (9781848320154)
—*British Cruisers* (978184832 0789)
—*Post-War Naval Revolution* (9780851774145)
Jordan, John *Soviet Warships* (9781854091178)
Jordan & Moulin *French Cruisers* (9781848321335)
Poolman, Kenneth *Allied Escort Carriers of WWII* (9780713712216)
Raven & Roberts *British Battleships of WWII* (9780870218170)
—*British Cruisers of WWII* (9780870219221)
Roberts, John *British Warships of the Second World War* (9781861761316)
Whitley, M.J. *Cruisers of World War Two* (9781860198748)
—*Destroyers of World War Two* (9781854095213)

SUPPLIERS

With the loss of local hobby stores that used to feature in most UK towns, many people have to buy the materials to build their models and equipment to outfit them from Internet sites. There are many businesses offering this service and space prevents all of them being listed.

As a starting point, the following are a few of the commercial firms that I have found to be useful for the materials needed to build working models in the UK. These will at least give you guidance about what is available, but a general Internet search can often locate the perfect item needed.

Wood – Balsa, Plywood and Lite Ply
SLEC Unit 8–10 Norwich Rd Industrial Est, Watton, Norfolk IP25 6DR
(sales@slecuk.com)

Electric Motors
MFA/COMODRILLS Felderland Lane, Worth, Deal, Kent CT14 0BT
(info@mfacomo.com)

Electrical Items
COMPONENT SHOP 1 Llwyn Bleddyn, Llanllechid, Bangor LL57 3EF
(infor@componentshop.co.uk)

General Modelling Items
HOWES MODELS Unit 16b Station Field Industrial Est, Kidlington, Oxon OX5 1JD
(radiocontrol@howesmodels.co.uk)

CORNWALL MODEL BOATS LTD Unit 3B, Highfield Industrial Est, Camelford, Cornwall PL32 9RA
(sales@cornwallmodelboats.co.uk)

Index

First published in 2024 by
The Crowood Press Ltd
Ramsbury, Marlborough
Wiltshire SN8 2HR

enquiries@crowood.com
www.crowood.com

**British Library Cataloguing-in-Publication
Data**
A catalogue record for this book is available from
the British Library.

ISBN 978 0 7198 4391 4

Typeset by Chennai Publishing Services
Cover design by Bluegecko
Printed and bound in India by Thomson Press India Ltd

Dedication

To my wife Susan, without her support,
tolerance and sometimes forgiveness,
I would never have gained the experience
to write this book.